THE EMERGING NEW SCIENCE

THE EMERGING
NEW SCIENCE

NOEL HUNTLEY

Copyright © 2019 by Noel Huntley.

ISBN: Softcover   978-1-7960-4295-5
        eBook   978-1-7960-4294-8

All rights reserved. No part of this book may be reproduced or transmitted in any form or by any means, electronic or mechanical, including photocopying, recording, or by any information storage and retrieval system, without permission in writing from the copyright owner.

Any people depicted in stock imagery provided by Getty Images are models, and such images are being used for illustrative purposes only.
Certain stock imagery © Getty Images.

Print information available on the last page.

Rev. date: 06/25/2019

To order additional copies of this book, contact:
Xlibris
1-888-795-4274
www.Xlibris.com
Orders@Xlibris.com
798437

TABLE OF CONTENTS

PART ONE
THE NEW SCIENCE
1. Introduction to the New Science   11
2. Is Science Really Progressing?   13
3. Science Problems   15
4. Fundamental Difference between Orthodox Science and the New Science   17
5. A Glimpse into the New Science   23
6. Features of the New Science   30
7. Orthodox Science and its Limitations   35
8. Interconnectivity of the Multiverse   39
9. Orthodox and New Science Compared   44
10. Quantum Physics   49
11. Harmonic and Non-Harmonic Sciences   54
12. The Mind, Chakra System and DNA   56

PART TWO
THE FRACTAL TREE
13. Science and Religion   65
14. The Evolution of the Species   67
15. Science and Technology   69
16. The Infinite Energy Source   71
17. Ego/God Dichotomy   75
18. Consciousness and the Paranormal   77
19. Ascension   79

PART THREE
EDUCATIONAL FALSEHOODS
20. There is No Life After the Death of the Physical Body   85
21. Scientific Methodology is the Only Acceptable Method of Acquiring Truth   91

22.	The Origin of the Universe: Big Bang Theory	94
23.	The Origin of the Species: Theory of Evolution	102
24.	The Mind is a By-product of the Brain	111

PART FOUR
THE NATURE OF CREATION

25.	The Nature of the Universe, Life and Evolution, and Their Purpose	121
26.	Dualism and the Anthropomorphic Principle	133
27.	Creation From the More Scientific Viewpoint	139
28.	Computer Systems	155
29.	Harmonic and Non-Harmonic Technologies	160

PART FIVE
THE PARADOX OF SOMETHING AND NOTHING

30.	First Cause and Infinite Regression	167
31.	Bringing in Quantum Computing	176

PART SIX
QUANTUM REALM

32.	Introduction	193
33.	Quantum Reduction and Quantum Regeneration	198
34.	Resolution of the Great Einstein-Bohr Debate	204
35.	Secondary Quantum Reductions	210

PART SEVEN
COLLAPSE OF THE WAVE FUNCTION

36.	Learning-Pattern Application	219
37.	Creating (Selecting) One's Own Reality?	232
38.	Conclusions	243

PART EIGHT
FURTHER RELATED TOPICS
 39. Quantum Teleportation   251
 40. Free Will   257
 41. Geometric Intelligence   270

APPENDICES

 42. APPENDIX A: The Solar System   281
 43. APPENDIX B: Conflicting Creations   284
 44. APPENDIX C: The Subjective/Objective
    Illusion   288
 45. APPENDIX D: Existence and Evolution,
    Quantum Reduction and Regeneration 291
 46. APPENDIX E: Overcoming the Velocity-
    of-Light Limitation   295

BIBLIOGRAPHY   301

INDEX   305

LIST OF ILLUSTRATIONS

Note: The illustrations were originally in colour (for the e-book) and there is some loss of information in the conversion to gray-scale. The coloured figures are available (free) on the website: www.nhbeyondduality.org.uk.

FIGURE 1	Universe model old versus new	18
FIGURE 2	Flatland analogy	19
FIGURE 3	Vortex	36
FIGURE 4	Universal vortices	42
FIGURE 5	Propulsion systems	55
FIGURE 6	Chakras	57
FIGURE 7	DNA	58
FIGURE 8	Fractal tree	64
FIGURE 9	Electrical power analogy	72
FIGURE 10	The ascension big picture	107
FIGURE 11	Introducing limitations	140
FIGURE 12	Fractal waves	147
FIGURE 13	Quantum regeneration versus emergent software	156
FIGURE 14	Triad principle	187
FIGURE 15	Degrees of order and complexity	199
FIGURE 16	Order and wave patterns	202
FIGURE 17	Einstein/Bohr debate resolution	206
FIGURE 18	Learning-pattern mechanisms	213
FIGURE 19	Waveforms	221
FIGURE 20	Program matrix	222
FIGURE 21	Learning-pattern example	224
FIGURE 22	Basic holographic principle: spheres within spheres	241
FIGURE 23	Arm vortices	275
FIGURE 24	Twelve-planet solar system	281
FIGURE 25	Creation concepts	289
FIGURE 26	Quantum realm and wave collapse	292

PART ONE

THE NEW SCIENCE

1.

INTRODUCTION TO THE NEW SCIENCE

'... we have to find a new view of the world'. Richard Feynman.

Mainstream science is encountering insurmountable hurdles in its struggle to unravel the many mysteries of existence and the universe. These difficulties are, however, essentially not given publicity by simply closing down the reach of science into a narrower context, and then with the assistance of the educational system and media eliminate or deny all phenomena beyond this contextual egotistically-limited boundary. This downgrading psychological 'procedure' results in outrageously premature solutions to the great questions of life and the universe. From this can arise scientific convictions that are sheer beliefs. People should be outraged at what is being thrust upon them through the media, in particular, television, moulding with total conviction the vulnerable minds of the young, who, with hardly an exception, carry these 'scientific' conclusions into adulthood, and continue to disseminate the dogmas. The answers to the great questions of life and existence are known today but not remotely through mainstream science.

Scientific data, measured with rigorous empiricism is one thing (and science is good at this, producing certainty by repetition of experiments), but scientific interpretation is another. Wishful thinking based on the needed intellectual security will draw conclusions that eliminate genuine unknowns and mysteries, producing that comfortable ego-state of mind free from confusions.

Let us list more specifically some of the justifications for formulating a new framework for science: long-standing theories of leading visionary physicists; quantum physics itself; New-Age movement; ancient wisdom; all the features of life and existence that mainstream science can't handle; paraphysics; religion; human origins; universe origin; the increasing chaos on the planet; science's inability to handle the concept of unity; holistic systems; inner space (and the 'within' of sacred teachings) and non-quantifiable aspects of existence. People should be horrified at what is being fed to them through the television, particularly its influence on children.

One might consider that the New Science already exists in bits and pieces throughout the world today and in history, even some large pieces that can be brought together—though a huge and difficult task (incurring great resistance from authorities in doing so). Scientists could pursue this task but won't, due to a combination of numerous reasons, such as being already educated (programmed) to consider there is nothing worthwhile outside established science. This is a form of arrogance and insecurity. But there is also intelligent manipulation of knowledge on the planet, which we shall not pursue.

However, it may be easy to point out the limitations of a subject, but can we justify introducing the new knowledge structures? Firstly let us briefly point out the importance of science, in general. Our lives revolve very much around science, and science revolves around the experimental method. However, the experimental results depend on the observer/observed relationship, which is not remotely understood (and is simply ignored). The basic science is physics, which underlies all other sciences, and physics is about how energy works—that is, how anything works. It thus has the potential to solve all problems on the planet and settle almost all conflicts and arguments.

Thus it is vital to the evolution of the civilisation that the observer/observed relationship is understood (see Part Three).

2.
IS SCIENCE REALLY PROGRESSING?

The paradox of progress

Science's true role is surely to embrace all facets of life, mind, universe, existence, the paranormal, and religion. But it fails utterly, except in the classification of phenomena of 3D matter, space and time, or, in other words, the grossly quantifiable aspects of life and physical expression, the particle level and its associated field systems—in other words, only the gross energies in an actually very sophisticated universe.

Classical physics, generally referred to as Newtonian, gave scientists and the public, who would accept it, a clear-cut deterministic model of the universe, which placed religion on shaky grounds. This standpoint was then strengthened by Darwin's theory of evolution.

Overall, a science based on a materialistic approach only, gives us a causation that proceeds from parts to wholes. The parts determine the whole, and a consequence of this is a cosmology in which a universe must begin with a random condition, and similarly for life. The essential problem here is 1) the refusal to look at the bigger picture, and 2) resisting an expansion of knowledge, preventing a recognition of the fact that established laws of science are relative; and opposing anything that requires an adjustment to existing established laws.

Let us bring home what we mean by 'expansion' with a simple analogy—a technological one. We can use the familiar example of the motor-vehicle combustion engine. The early inventor of this engine clearly had a remarkable faith and insight into his idea that a series of destructive explosions could effectively drive a wheel—he must certainly have been ridiculed (even more so by the educated of those days). Today we know that we can sit in comfort in a motor vehicle, which has a silent engine and finger-tip control. This is taken for granted now but it demonstrates the remarkable ingenuity of engineers and scientists. Unfortunately no matter how perfect it is, it is the same old crude system of locomotion: generating huge forces and inertia, with severe friction factors, ludicrous inefficiencies, and not to mention resulting pollution.

What this means is that knowledge is being consolidated, refined and perfected, not expanded into new territories of discoveries, new principles, new inventions. This is not real progress, which should be expansion into new fields and new laws; not improving the same old data.[1] [However, in this particular example, to be fair, we know it is not entirely the limited thinking of the inventors but also suppression of new technologies.][2]

3.

SCIENCE PROBLEMS

Science is burdened by the following problems:

It does not provide an acceptable beginning to creation and explain the basis of religious knowledge.

It does not explain all the dark matter of the universe.

It does not explain the order in nature and the universe.

There is no science of mind (or life).

There is no explanation for consciousness.

It does not recognise and cannot handle the concept of continuous existence.

It does not recognise and explain obvious instances of the limitations of Newton's laws, and the limited applicability of most science laws (that they are only correct relative to their own context).

There is no understanding of the observer/observed relationship upon which the whole edifice of science and knowledge is founded—exposed by quantum physics.

It does not have an understanding of the subjective/objective relationship, nor the absolute/relative relationship, and the basis to all knowledge and energy that they are contextual. *All energy and knowledge are contextual.*

It has no explanation for the 95% so-called 'junk' DNA.

There is no satisfactory explanation for the origins of man and the nature of evolution.

It propounds the irrefutability of the velocity of light limitation as applied to all bodies.

There is no recognition of the significance of zero (the most important number and is relative) or the universal geometric number 12 (and that the universe operates on geometric intelligence).

Science unwittingly quantum reduces truth to lower orders.

The ubiquitous fractal is a mystery to science.

It does not understand the part and whole relationship.

4.

FUNDAMENTAL DIFFERENCE BETWEEN ORTHODOX SCIENCE AND THE NEW SCIENCE

The difference between constructs directed from the part or from the whole.

To give the reader a quick visual reference, even at this early stage, and clarify a principal basic difference between the structural scopes of conventional science compared with the New Science, take a close look at Figure 1.

In Figure 1(a) we see in a simplistic presentation the type of framework orthodox science gives to the structure of the universe. There is just one level; that of particles, or particles stuck together by forces, forming groups, such as molecules, planets, universes. By 'one level' we mean 3D, which science considers is sufficient to handle the macro-universe.

Now look at Figure 1(b) showing the structures from the New Science's viewpoint. The difference might appear unbelievable at first. This is a representation of the holographic, fractal structure. There is organisation within organisation, interconnected on different levels of dimensions and frequencies. The linear, simplistic structure (the 'surface' visibility) of (a) is missing the internal structure shown in (b). Additional laws and principles of physics are required, such as, in natural bodies, underlying the surface particle level there are undivided wholes of energy, and the larger the (true) whole—a quantum state—the higher its frequencies. (Note that leading quantum physicist David Bohm theorised the existence of these larger quantum states, from particles to greater entities, such as a planet, to even the whole universe more than thirty years ago[3].)

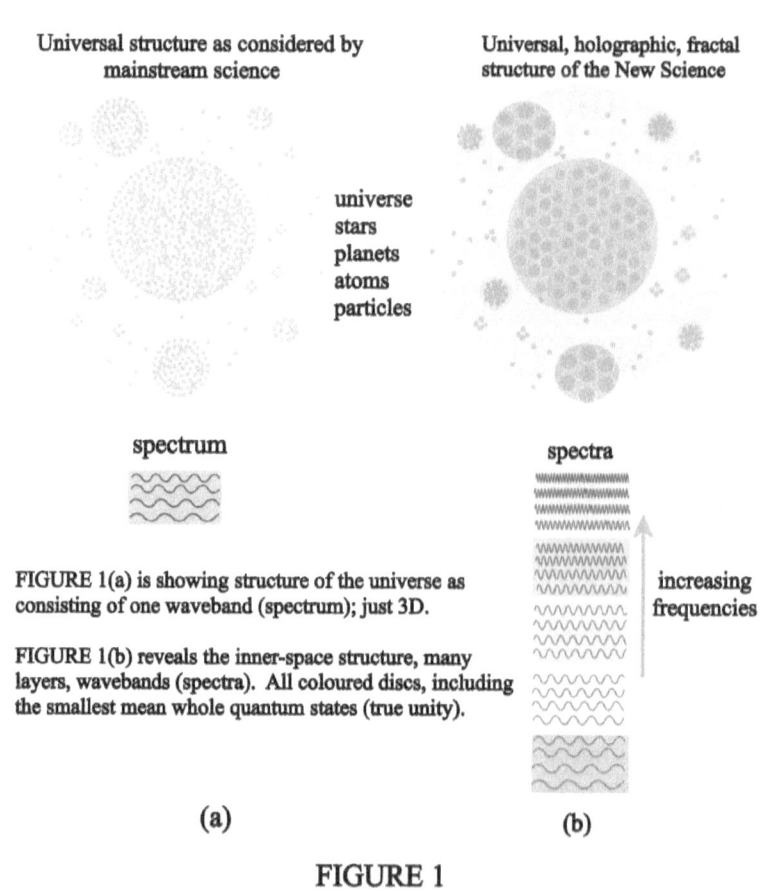

FIGURE 1(a) is showing structure of the universe as consisting of one waveband (spectrum); just 3D.

FIGURE 1(b) reveals the inner-space structure, many layers, wavebands (spectra). All coloured discs, including the smallest mean whole quantum states (true unity).

(a)   (b)

FIGURE 1

The larger wholes (note the colour scheme) can be seen to be higher in the spectrum than the smaller wholes. The inner (internal) space is diagrammed as higher. This inner direction aligns with the 'within' stated by the great spiritual leader, Jesus Christ, 2000 years ago ('within' referring to the location of Source: 'Kingdom within,' 'Father within'—the clue to the reconciliation of science and religion). Hopefully this will be clarified sufficiently for the reader later in the section on the Fractal Tree.

How can one visualise this inner space in a simple diagram? Where are the higher spectra of Figure 1(b) in relation to the one-level frequency band of Figure 1(a)? [Note that we shall generally use the abbreviations 3D, 4D to designate three-dimensions and four-dimensions, respectively, which emphasise dimensional geometry, but if, say, D3 is used instead of 3D, we are emphasising dimensional spectra.]

It may be helpful to recall the 'flatland' analogy. In Figure 2(a) we have a 2D surface with a circle or disc marked on it. We imagine 2D flat beings perceiving this. As they walk around it, it would appear as a closed sphere does to us. Now in Figure 2(b) we see that the cylinder impresses a disc shape on the 3D

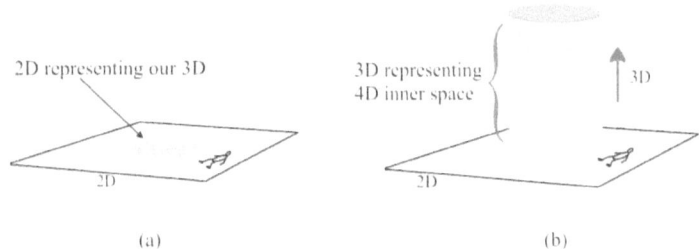

FIGURE 2

surface, and the flat beings would still see the disc as before and appearing like a sphere does to us. Thus the disc in (b) extends into the inner space (going up in the diagram). It has a thickness in 3D (the height of the cylinder), corresponding to a 'thickness' in 4D for us, which we can't perceive.

In Figure 1 we have a similar situation and correspondence between (a) and (b) in both figures.

Many of these ideas are not new to the visionary scientists. Let us quote some of the leading physicists. Professor John Wheeler of Princeton states that the geometrodynamical quantum foam of superspace represents a superhologram of the universe. Physicist Jack Sarfatti's interpretation of this quantum superspace is that the wormholes connect all parts of the universe directly to every other part. Leading quantum physicist David Bohm stresses quantum interconnectedness and unbroken wholeness. Charles Muses and Arthur Young refer to objects as superhologram images. Physicist Keith Floyd states that holographic models of consciousness make such brain processes as memory, perception, and imaging clearly explainable.

Renowned philosopher and mathematician, P.D. Ouspensky, in his book *A New Model of the Universe*, states that everything is everywhere and always—a description of the holograph in space and time. Also Philosopher Leibnitz's brilliant analysis of unity requires the holographic interpretation to understand it, which was unknown in the 1700s.

Science writers Michael Talbot reiterates that thought processes are holographic in that all thoughts are infinitely cross-referenced with all other thoughts, and Fritjof Capra speaks of the universe as a hologram, in which each part determines the whole.

We might be familiar with the philosophical statement, 'As above, so below', which is based on the axiom of Hermeticism, 'What is here is everywhere; what is not here is nowhere'. This again is the holographic property.

Restricted and controlled science, tailored specially for the public, ignores all this. Any knowledge that takes one beyond the boundaries of a simplistic random universe is suppressed. Academics need to recognise that the universe or multiverse is far more complex and intelligently designed than is supposed.

Science has created its own impassable boundaries and set its own limits of knowledge due to the belief that the only

The Emerging New Science

acceptable way of acquiring truth is through the experimental method, or so-called scientific methodology. In effect, science basically puts the lid on the third dimension and denies anything beyond this. A transition is desperately needed to take us beyond this barrier but which cannot be achieved by the current denial that all knowledge acquired by means other than the experimental method is not acceptable.[4] What is the solution?

Over the years the basis for a New Science has been emerging. It is not generally realised and certainly not taught on this planet that in addition to orthodox science, which is taught or communicated through the educational system, scientific sources, television and the media in general, another category of knowledge has gradually surfaced, which as we organise it, it becomes another body of scientific data (but relatively unorganised and untested by the scientific method). Much of these more scattered areas of knowledge can be found to have common ground, in particular, at a basic universal level, and the nature and origin of life. This material is an invitation to academics to extend their thinking into these, at present, peripheral areas of science.

A drastic change from the current orthodox paradigm is vital to the proper survival of the species; advice heralded by many leading physicists. Science's modus operandi embraces far too narrow a framework to accommodate the complexities of life, consciousness, and even the nature of the universe. Note leading physicist Richard Feynman's comment: 'What we need is imagination—we have to find a new view of the world.'

Thus today this second category of knowledge is emerging into public awareness—involving new thinking, new knowledge and scientific information, which we are designating the New Science. There is currently a failure to recognise that we have inherently these two categories of science since they are apparently of very different viewpoints. Each is looking at the universe in a different way. This automatically gives rise to receiving different information from the universe: the type of observation made determines what is 'seen'. Quantum physics exposed this years ago but its massive importance has been ignored—see Part Three, Section 2.

We shall see that in fact there is a bridge that can take us from the old to the new—which includes concepts already accepted by many scientists—and it is quantum physics, fully recognised within modern science.

Nevertheless note that quantum physics has come under more attack, disbelief and (past) rejection than any other discipline of science (even with experiments or mathematics to back it up). The existing framework of science prior to quantum physics couldn't accommodate it, in particular, the 'bizarre' notions of this revolutionary way of thinking. The elements of the New Science similarly come under attack since they don't yet fit into the old science.

5.

A GLIMPSE INTO THE NEW SCIENCE

Recognition of higher orders of knowledge

Regarding the recognition and acceptance of the New Science, owing to its lack of apparent order, consistency, integration and fragmented appearance, it wouldn't be viewed as a single subject, in particular, by the academia, unfamiliar with this seemingly diverse, unauthenticated and conjectural information. Many individual researchers or independent groups are in fact working parallel on similar projects, and many new perceptions, conclusions, inventions begin to align. This process of integration is occurring both tangibly between such parties and also in the subsequent observational correlations and conclusions. Thus there is an incremental cohesion taking place of previously separate, disparate areas of research, with increasing degrees of agreement, moving towards a consistent body of knowledge, which although initially will inevitably contain degrees of speculation, can conceptually dwarf mainstream's experimentally-dependent sciences, as the new knowledge reaches into all facets of existence and experience.

Let us present a few examples, merely to show how much more can be known today on the planet using other sources than the experimental method. At this point we are not endeavouring to convince the reader of the validity of these examples; the purpose is merely to illustrate that there is information on this planet that is available to be known at this time, which is thousands of years ahead of current orthodox science. We are trying to show the importance and value that this can be to conventional science. If there is a very persuasive element of truth in something of this magnitude, shouldn't science follow up on the information and source—check it out?

Unfortunately it doesn't occur. Academics will scoff at it with the usual arrogance, even be outraged with great confidence at the audacity of anyone making such claims without the experimental proof. Note that under these circumstances opposition to such information will be based on anything from a disturbance to intellectual security, arrogance, a context-dependent thinking (anything new then doesn't 'resonate'), to a programmed prejudice against certain unorthodox sources.

These examples will really test the belief of the reader (which means the fixed learning structures in the mind dictating what to think). However, at this stage we are only showing how unofficial data (relative to rigorous scientific methodology) may present very valuable insights that could be pursued to great advantage.

The first example is one which science wouldn't have 'proof' of in 10,000 years or so of our planetary evolution of knowledge. Our Milky Way galaxy, consisting of hundreds of billions of star systems, has been scientifically established as containing a black-hole at its centre. This is considered to be normal for a galaxy. But in fact it is a degenerating galaxy; the opposite of a normal, evolving galaxy that has a black-hole/-white-hole in balance (the oscillations). Our black-hole galaxy has a predominant black-hole and is slowly compacting, imploding, taking millions of years to go through this cycle. However, it doesn't all have to degrade (this is a variable).

Some hundreds of billions of years ago our Milky Way galaxy separated away from the host or parent galaxy of Andromeda (which contains over one trillion stars compared with ours, some 300 billion stars). We can explain later how science can arrive at such an underestimate of the age of the universe as

The Emerging New Science

some 14 billion years, compared with hundreds of billions as implied above.

Keep in mind established science is only looking at the 'surface' of the universe and existence. The low resolution of scientific instruments cannot detect the higher frequencies within inner space, nor can they detect objects and particles when oriented differently in inner space (we shall give attention to this later). Keeping in mind that everything is structured from frequency patterns based on frequency spectra, a degeneration took place in regions of Andromeda, resulting in a lowering of frequency of the corresponding spectra. For evolving Andromeda, this is counter-evolutionary, causing a degenerating influence on the galaxy. However, this was accompanied by a shift in orientation of the particles collectively of this regressed energy mass (nascent Milky Way). It stabilised at an angle of 15 degrees to Andromeda (and with respect to its original mass/-energy). This orientation is within inner space, that is, in a 4D direction. This rotation within inner space then corresponds and creates a projected linear difference in 3D space of about 2.4 million light years (an amazing concept). And today we have an apparently separate galaxy, the Milky Way, isolated from its origin, the Andromeda galaxy—exceedingly separate in 3D space but 'adjacent' in inner space. Regions of our galaxy that are not declining will return to Andromeda, as in fact speculated by orthodox science (but returning not in the same manner).

We know in orthodox science about wormholes (this term came from physics, not science fiction), and it is speculated in science that in entering a wormhole one will encounter changes in space-time and could feasibly finish up at a far-distant exit point back into our 3D in a very short time—a little like imagining folded space (and taking a shortcut at the point of the fold).

The final picture then is of our Milky Way, 2.5 million light years away from the Andromeda galaxy in outer space, but within inner space it is, in effect, located directly 'below' Andromeda with a common point of contact still connected directly—the pivot point of the 15-degree orientation—and amazingly the region of Earth's location.

Is this a return of the original false notion of Earth being the centre of the universe (then considered the solar system)? No, but Earth is nevertheless within this region of space-time—the common axis of two galaxies (this has huge implications).[5]

The next example will be an even greater jolt to the closed, secure mind of any such reader, in particular, it might be a fearful piece of information. Again, and if necessary, take on the neutral state of mind, neither believing nor disbelieving, but use it as an example, at least initially, in the category of 'what if it were true'. This example clarifies the mystery and confusion around the following significant phenomena: Wormwood ('planet' mentioned in the Bible), planet X, Nibiru, Stonehenge, the mysterious asteroid belt, and the 'junk' DNA.

The event which brought together the above components was passed down in human religious terms as the Luciferian (or Luciferic) rebellion. This apparently occurred around 25,500 BC. The 'planet' Wormwood was operated as a relay station by the large (Anunnaki-Resistance) inhabited planet Nibiru, which is currently approximately at its furthest point from Earth (in the Pleiades system) in a huge elliptical orbit around our Sun of period 3,657.8 years. Wormwood was a large fragment from a planet called Maldek, of which the latter was destroyed, leaving the asteroid belt—see Appendix A.

Thus the Nibiru race commandeered the planet Wormwood and used it to communicate to and interface with Earth using scalar-sonic pulses to activate a massive crystal (Nibiru diodic crystal), which had been inserted deep within Earth, under where Stonehenge now exists (the crystal would have been embedded astrally, via the 4D spectrum). The crystal then emitted programmed frequency patterns that spiralled upwards to the surface of Earth. Then as the energy rotated, it projected out 16 scalar beams across most of the globe. This procedure was directed at and subsequently interfaced with the DNA of all life, causing most of the so-called 'junk' DNA (disconnecting it and unplugging those regions, which previously had base pairs, resulting in loss of these molecules). Its main purpose was to erase cellular memory of all life on planet Earth (which it did). People do not realise it is not normal to not remember past lives and even the beginnings of the civilisation, since all the information is there, available in every cell. (Remember, history has been re-written, very successfully.)

Now, we might anticipate where Stonehenge enters into the puzzle. Humans were advanced at the time and were in continuous communication with visiting off-planet races, in particular, the Sirian Council and benevolent Anunnaki representatives from Sirius. The intruder Anunnaki resistance from Nibiru were kept in check by the Sirian Council. (The history of this is given and how it relates to the Great Pyramid in the book The Original Great Pyramid and Future Science.)[6]

Therefore, subsequently the Stonehenge 'monument' was built, which had a simple but precise geometry to counteract the programming. The spiralling energy from the crystal flowed up the stone pillars, circulated around the ring of stone slabs on top, then was reflected back down again, preventing the detrimental distribution of scalar programming, manipulating the civilisation and life in general. There is much more to follow up on here thus let us finalise this before continuing with the third example.

Wormwood was prophesied in the Bible to interact with Earth in our period of time and has been, and still is, referred to as Planet X and publicised on the Internet. The prophecy was essentially correct, except that the predictions are due to cycles and planned agendas. Wormwood was due to swing round between the Sun and Earth on May 27, 2003, causing pole shift.

On the Internet, the Zeta channellings on their website made it clear that this would occur on this date, causing the destruction of the civilisation. Contrarily, other groups knew about the promised intervention from the Guardian races, who fortunately executed their plan A and successfully diverted the catastrophe. However, the first part of Plan A only succeeded in reducing the size of Wormwood by about 20%, possibly leaving some asteroids. The second part of Plan A wasn't made clear for many months as it involved something well beyond our science and technologies, even conceptually, but which involved shifting Wormwood into a different probability time line.[7] Plan B wasn't required, which was a rather desperate measure, in which the Guardian race would have landed on Earth and secretly worked with government groups to physically handle this threatening body, incurring physical attacks from the intruder extraterrestrials.

The third example is a very important one for everyone to know about as it is instrumental in determining anyone's future progress after death of the physical body. This is the subject of whether to bury or cremate. Our New Science obviously recognises the 'soul', which is simply the next higher aspect of consciousness, or next higher fractal level (explained later). There is no mystery about the soul (or whatever one wants to call it). The soul body has more complex physics than the physical body.

Now when the body dies we envisage the soul body leaving the physical body. However, there is a huge difference between their spectra of frequencies. The soul is structured from a higher spectrum with higher frequencies, clearly evidenced by the fact that we can see the physical body but not the soul body. Even more emphatic is that scientific instruments can't detect the soul body, owing to their relatively low resolution (like a camera with film of low resolution that doesn't pick up much).

This also means that the two bodies are incompatible. The higher frequencies of the soul will not lock into the low frequencies of physical body's atomic structure. An etheric interface body is required, just as we require, for example, an interface to make compatible two dissimilar pieces of computer hardware/software. The interface body thus enables the soul to resonate with the physical body and hold the two together.

Unfortunately at death of the physical body the interface remains with the physical body. It should go with the soul, since it is needed in the next existence and necessary for the soul personality to resume its evolution/ascension. The soul loses a portion of its own consciousness energy that was used when incarnating to provide body consciousness. The mechanism is that distortions in the physical body due to certain mutations in all DNAs on the planet cause corresponding impairments in the interface body and it can't detach itself. The interface body then resonates more with the physical body than the soul body. This causes the soul body (the 'person') to incarnate back into that approximate space-time location, repeating similar lives with no real progress. The simple answer is cremation. This immediately enables the interface body to return to the soul. (One will never find physical body remains of advanced civilisations, all of whom would have cremated or used vaporisation techniques.)

6.

FEATURES OF THE NEW SCIENCE

A 'top-down' paradigm as opposed to 'bottom up'.

The New Science would have the potential to redirect civilisation and restore its natural evolutionary heritage. This New Science will handle the missing pieces of the old science: the vast array of universal phenomena that scientific methodology cannot detect. However, orthodox, mainstream science, as stated, is ruled and hugely handicapped by the tenet that the only acceptable way of obtaining truth is through the experimental system. Consequently on this (erroneous) basis it would not at this time entertain the validity of the New Science. Nevertheless, nothing can stop the new knowledge from evolving (unless minds become sufficiently robotised). These two sciences, however, mainstream and the new, can complement one another if the old can recognise that the experimental method is limited and then realise the subsequent ramifications of this, such as having only relative laws; the physical constants of science are not truly constant (only over a certain contextual range). Evidence is within quantum physics, explained later.

Many basic truths (which are, however, often actually speculations and theories) taught through the media, education and science, are about as opposite to truth as they could be. Let us present a few examples. A major one that just about everyone is familiar with is the scientific belief that life comes from matter, rather than the other way round, that matter is a derivative of life. The higher-order systems of life cannot develop from the lower order of matter, though scientists are desperately trying to prove this (see later discussions). An extension of this is that energies of mind and consciousness are by-products of the brain (which is matter), compared with the New Science, which recognises that brain and matter, in general, are by-products of consciousness; a viewpoint held by many leading scientists, in particular, quantum physicists.

Another example is that the universe came into existence from nothing (Big Bang theory) and from a random condition. Science desperately needs to at least postulate a beginning state and recognise that ordered directions began the process of evolution of all natural systems, atoms, molecules, cells, planets, stars, galaxies, universes. The old science tells us that evolution has a direction of causation that is only 'bottom-up' from particles, or maximum fragmentation to larger forms, rather than first cause being primarily 'top-down' from greater wholes (of consciousness) to lesser wholes, with lesser wholes returning to greater wholes in the evolutionary process.

Then of course we have the classic one of man from apes, rather than in fact a degeneration of the human race from a higher state of 'evolution'—this should be elementary information and is, if one is sufficiently unbiased and prepared to look at the new information. Another example is 'survival of the fittest', which is emphasis on self, the ego (a certain path for ultimately the destruction of the human race); compared with integration and one's relationship and responsibility to the whole.

A final example is the blank region, which is over 95%, of the so-called 'junk' DNA where base-pair molecules have broken off from a much more advanced DNA. Its unplugged higher-spectra strands remain undetected by science; see Figure 7.

The New Science recognises that the universe operates on the geometry of 12, that is, on geometric intelligence. Orthodox science merely recognises that geometry is the basic language of the universe, but not realising it is based on 12 (mathematics needs to change its base from decimal (10) to twelve).

What are the sources of the New Science? Firstly it utilises the tried and tested concepts and theories of established science. Fundamental terms of orthodox science are now established, for example, force, mass, energy, space, time, gravity, fractals, holographic, quanta, electromagnetism, which are extended in scope by the New Science and, in particular, include a proper recognition of underlying coherent states to bring an expansion of science throughout many dimensions of existence. Some features that are lacking in the old system but researched in the New Science, would be emphasis on the universality of frequency patterns, templates, vortices, quantum regeneration, mind, consciousness, antigravity, negative resistance, over-unity systems (true macro-), dimensions, and more references to fractal systems, in particular, internal ones, and the term holographic (both in space and time). Information is obtained from many diverse sources, visionary physicists, metaphysics and 'lost' ancient knowledge, psychology, parapsychology, religion, philosophy, New-Age and extraterrestrial information/trans-missions, and new theories by many independent thinkers. Intuition, in particular, is a vital faculty, though being generally very weak for humans on this planet it is a real bonus when it is functioning. It is not properly recognised within mainstream teachings and totally underestimated. Thus intuition is not validated by science, even though it is fully capable, if sufficiently developed, of identifying truth, even where structural knowledge is absent.

Physical empiricism (observation and experiment) could ultimately be replaced by mental empiricism, in which the trained individual through developed faculties, such as the intuition, can perceive all as mind and to separate from a region to observe it, creating a sufficient degree of objectivity to evaluate results. An advanced civilisation does not become involved with the limitations of physical experimentation.

Specific features of the New Science:
- We are in a multiverse of many universes harmonically designed in a fractal holographic configuration. It

is as much a work of art as a great machine or computer.
- The dual counter-rotating vortex is the basic energy configuration, which circulates the energy throughout the fractal holographic multiverse.
- All energy and knowledge are contextual. Everything observed, detected, evaluated, is in the context of whom or what is doing the observing and from what perspective or station.
- The fractal law. One of the few truly universal laws compared with science's relative laws. A basic model for the New Science would be the fractal tree. Each fractal level, for example, twig, branch, larger branch, etc. is structured from a spectrum (or waveband) of frequencies and dimensions—also has its own vortex quantum state. (See Part Two, The Fractal Tree.) Each dimension has 12 subharmonic frequencies and their rates of vibration increase from the first dimension upwards.
- The New-Age field generally recognises the existence of the multiverse and its multidimensionality with frequency spectra at each dimensional level but obviously has no ordered and consistent science.
- The multiverse functions on base 12 and is an ordered structure of dimensions specially evolved from blueprint templates for the exploration of consciousness, which exists on all dimensional levels. Our lowest 3D universe, in general, has the greatest limitations, with limited degrees of freedom as we evolve (ascend) through this hierarchical structure, with changing bodies.
- The physical body and genetics are merely a vehicle serving the purpose as a template for the unlimited nature of consciousness to focus (and be formatted) within and to explore that particular dimensional existence. Consciousness is a complex structure. It

records all perceptions and actions forming a mind, which it closely unites with, indistinguishably, to most individuals. However, underlying all this mind structure is a state of pure consciousness, nonquantifiable, innate native-state awareness, or sentience, beyond space and time. This alone gives the experiential (aliveness) aspect of life[8] (mind/body structures mould this sentience into patterns, frameworks, to modify the experience, such as create pain—not, however, part of a harmonic evolution). See Part 3 on continuous life.

- The mind is not a by-product of the brain, as orthodox science teaches. In fact, the mind precedes the brain and body. Further, consciousness is not a by-product of the brain or mind; consciousness precedes the mind. Consciousness precedes all matter, that is, matter is a derivative of consciousness (as stated very correctly by Schrodinger: leading quantum physicist responsible for the famous quantum mechanical wave equation).
- The old science recognises quantum reduction only; the New Science embraces both quantum reduction and quantum regeneration.
- In the New Science everything has its underlying wave pattern, which manifests what we see at the surface level, that is, our reality.

7.

ORTHODOX SCIENCE AND ITS LIMITATIONS

Scientific stumbling blocks

The desire within science to explain everything would appear to be greater than the experience of having the knowledge itself and its uses. Science, instead of moving out into new fields, concepts, greater laws, seemingly compulsively pulls those nascent regions into existing closed-minded paradigms, sometimes quite outrageously forcing them to fit. Quantum physics has indicated those new fields; is science taking advantage of this?

Unless scientists exercise some humility and give some attention to the New Science, which will continue to gain increasing attention and support from the public until academics are forced to consider this new view, it will begin to enforce its participation with orthodox science, in particular, as increasing numbers of academics covertly study, research and verify the new data (already occurring). We say 'covertly' since knowledge can be the most dangerous possession in the world and there are heretical terms and concepts in universities that can render professors fired. We are referring of course to the features of the New Science. (At a more fundamental level, beyond the scope of this book, such features are deliberately being suppressed, since they are known to be valid and not allowed to be made available to the public. Some heretical terms are, antigravity, over-unity energy generation, negative resistance, anything perpetual or continuous in open systems, and of course the well-known free-energy device claims.)

There are many groups throughout the world researching vortex theories. Academic research into this subject took an unfortunate turn in the early days of Lord Kelvin and was discontinued. They didn't postulate the dual counter-rotating vortex and take it to the next dimension (see Figure 3). This would have revealed instantly the much sought supersymmetry and the fundamental principle of all creation, the particle and antiparticle side of all natural bodies and entities (including universes).[9]

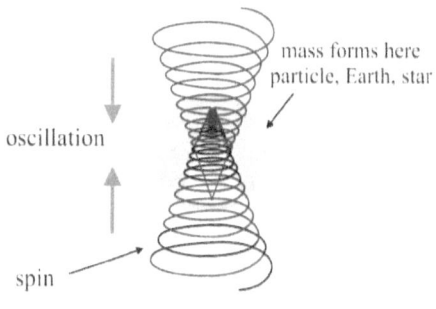

Dual-vortex polarity - 3D type
exists as above, but is also 4D

FIGURE 3

Mainstream science is limited to the 3D spectrum but all dimensions interpenetrate and we can tune into and utilise the 4D spectrum and higher, which are embraced by the New Science. For example, in researching the nature of skills, of which the basis is the learning pattern, conventional science limits itself to the brain and physiology. The New Science covers higher-dimensional spectra, the basis of mind, and the learning pattern can then be understood for what it is: a 4D holographic template that houses programs and converts nonlinear information into linear information. The learning pattern is holographic in time (short interval) as well as space. Current science can't remotely handle this with its emphasis on the brain.

Conventional science does not have a science of the mind. Physics is the basic science and should comprise all dimensional spectra (wave bands) and fractal levels. Physics is being manipulated against embracing energies above 3D. One way this manipulation is achieved is simply by stating there is no mind (that it is only a by-product of the brain). Thus, regarding the human being, the old science handles only the body and brain; the New Science embraces mind, brain and body, and the nature of consciousness, extending into the higher aspects (fractal levels) of consciousness, such as the so-called 'soul', and precisely what this is. The soul is no mystery in the new knowledge. It is in fact the second (dimensional) fractal level 'up'. Picture the fractal tree (explained later). Remove all leaves for clarity. Let the twig level correspond to our 3D human dimension; the branch it is attached to is now the soul level (the multiverse will also be fractalised).

Orthodox science believes in a universe made of matter, forces, and 3D space with the extra dimension time. Science has not yet recognised that these dimensions have frequency spectra. That is, each dimension is a frequency spectrum or waveband out of which our 3D reality is created. The reason science is not detecting these spectra is because 1) the observing part of the experimental method, which is the observer's physical senses, cannot supply information to make a proper judgement in the larger picture; it can only achieve relative values (to that context) its viewpoint is inside the context of that which it is evaluating, and 2) the scientific instruments, are also structured out of these same wavebands. Quantum physics indirectly exposed this more than fifty years ago, showing that the observer is *part of the experimental set-up*; vitally important information to which we shall return.

The New Science encompasses these dimensional wavebands and above 3D, in which the frequencies increase into higher dimensions, that is, rates of information. The soul level is within the next fractal universe system D4, D5, D6. Note that our 3D space has a low rate of information and the velocity of light is governed by this, but can be exceeded in harmonic higher-frequency systems. See Figure 5 and Appendix E.

The New Science recognises the best features of quantum physics, as mentioned: that all observations are contextual (in the context of who or what is doing the observing); all energy, such as radiation, is not continuous but composed of discrete, whole states, called quanta; non-locality (global effects) and superposition (see later sections). Quantum physics has come under more attack than any other scientific discipline—the more truth one puts out the more one will be attacked on this planet.

A summary of scientific limitations would be: 1) the contextual aspect of the scientific method, see Section 21; 2) experimental observations may quantum reduce the coherence of higher truths to lower-order laws, see later sections; 3) there are inherent within nature and the environment certain distortions, mutations, affecting the evaluation of laws, see Section 21; 4) adverse influences and control from 'covert politics' (not pursued here), 5) retarding effects from nonproductive psychological states, such as arrogance, lack of humility, intellectual insecurity, and so on.

8.

INTERCONNECTIVITY OF THE MULTIVERSE
Fractal, holographic systems

We have referred to the interconnectedness of the multiverse. For this, envisage everything one can know about, energies, objects, particles and life, as interconnected in a network of infinite energy lines filling all space, linking everything together (in the background or inner space or the 'within'). This is supported and suggested by quantum physics, and is elementary in non-mainstream sources that forms part of our New Science. Why is it infinitely connected? Because it is a product of consciousness and reflects its infinite non-linear nature—we shall return to this.

What kind of organisation has this interconnectedness? This multiverse is a fractal, holographic system. For those not familiar with fractals or the meaning of holographic, just think of the analogy of a company organisation in which its staff members are ranked in a pyramidal configuration (for example, ground-floor workers, managers, executives and president). This is a good simulation or analogy for the fractal, holographic systems (the president is connected through the network directly to each subpart, such as executives, managers, workers, and the fractal levels are the ranking levels).

Mainstream science is only viewing the surface of this multiverse of spinning celestial bodies with orbits within orbits (satellites orbiting planets, planets orbiting stars, stars orbiting larger stars, and so on); all claimed to be regulated and 'sculptured' essentially by gravity. It is totally absent of the inner-space machinery, which interconnects the many dimensions in this fractal, holographic design; gravity doesn't directly hold together the universe but is a by-product of the vortex.[10]

The key mechanism or energy unit of this inner-space machinery is the vortex, configured in a dual-polarity counter-rotational system. It is these electromagnetic forces of the vortices that are primary in holding together the universe, not gravity. Science cannot remotely detect these high frequencies (except the frequencies of atoms, particles). See Figure 4.

Even with a quantum theory of the atom, much superior to the early atomic models (for instance, the 'billiard-ball' configuration), there is still a higher order lacking; current models are three-dimensional. An atom would not hold together without the higher dimensions. The vortex model, Figure 3, is fundamental to all material natural objects, particles, planets, stars, galaxies. In the figure we can see that by bringing closer together the spiralling energies of the two poles they will create intersection points in the rotating, spinning, lines of force, these intersection points then become vortices themselves, that is, particles, such as electrons. The formation of orbitals would be a natural consequence of the interference pattern from 3D fringes creating nodes within alternate light (occupied) and dark (non-occupied) shells. The same principle would apply to planets and other bodies in orbits the orbiting body coherent (resonating) with a rotating node.

It is easy to see that the electrons have no separate existence (they could nevertheless radiate across space as waves and then be detected by a particle-detection scientific instrument). We also see that the atom is one whole even the two vortices come from a single state (the triad principle) and with a dominant inflow of 'aether' (quantum vacuum energy), creating a negative charge. This flow would circulate in 4D and out at the nucleus, containing protons, giving a positive charge (white-hole effect).

The vortex model also provides us with a more coherent structure for gravity. Einstein recognised the apparent absence of the force of gravity in free fall and thus is superior to Newton's interpretation. The principle of equivalence enabled Einstein to say that gravity is acceleration and introduce the geometry of curved space (4D curved) in his general theory of relativity for providing the effect of gravity. The theory posits that mass curves space and curved space creates gravity. This is two separate entities, mass and space. The vortex model not only provides the curvature but also the means for the formation of mass from the vortex; this also simultaneously creates the space (an oscillation from space to mass, biased towards the mass). And again we have a higher-order coherent level of organisation with the vortex model.

The interconnected vortices circulate the energy throughout the multiverse keeping it in continuous creation; a perpetual power house. Mainstream science denies and contradicts this and demands that there can only be closed systems with limited energy (as required by the electricity and power controllers). Science places us in a closed 3D box with fixed energy that is transformed from one type of energy to another with reducing energy potentials (reducing order), leading to the inactivity of an entropic death (no order, a steady state of equilibrium and inactivity).

As per the New Science, the energy is entering and leaving the system, say, 3D it is dynamic and perpetual. An atom is an energy unit; it is in continuous oscillation for a normal or original atom, one not part of a manipulated malfunctioning closed system. One could tap its energy by resonance of the oscillations and the atom would not become depleted. The old science is based on finite sources; the New Science is based on perpetual sources.

This continuous, perpetual circulation of energy throughout the multiverse, which has fractal and holographic order, is fully harmonic at a sufficiently basic level. However, as indicated above, just to complicate matters further, on the surface in local regions of this creation there are malfunctions (a form of mutation), in particular, in our Milky Way galaxy. We shall not pursue this other than present a brief reference in Appendix B.

If life, which has a degree of free will, damages any regions, such as causing harm to others (distorting the appropriate harmonic-frequency patterns), a natural universal law is activated (override systems, immune systems, 'antivirus programmes' or karma, whatever one wishes to call them). This attempts to re-balance, re-align the energy to restore harmony. We shall say more on this (karma) later.

It might be of interest to tie together the universe vortices in Figure 4 with the different view of the universe in Figure 1(b).

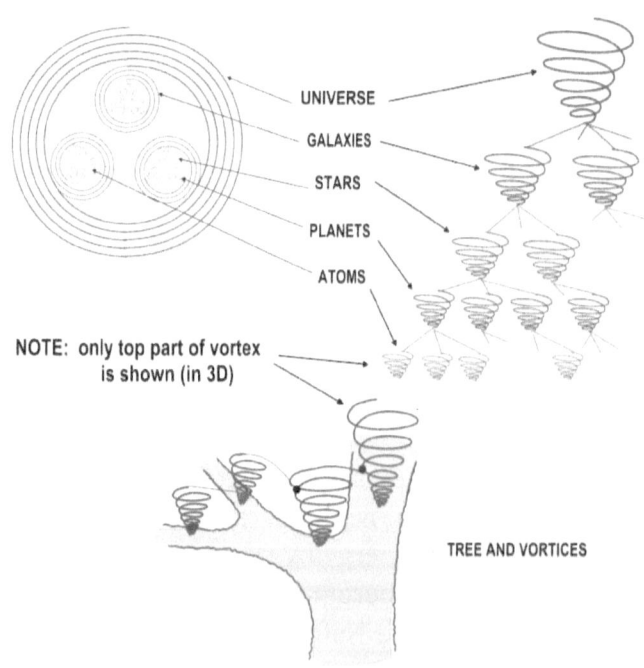

FIGURE 4

Each circle of colour is a single vortex polarity. The circles within circles are the same as the vortex network of the universe, stars, planets, atoms, shown in Figure 4.

9.

ORTHODOX AND NEW SCIENCE COMPARED

ORTHODOX, MAINSTREAM SCIENCE	THE NEW SCIENCE
Non-harmonic science and technologies.	Harmonic science and technologies.
Three-dimensional universe.	Multidimensional universes.
Evolution from a random condition and progressing by natural selection, and applying only to organic existence.	Evolution from ordered seeding imprints for all life and natural bodies through the dimensional structures of the multiverse.
Not handle synchronistic phenomena.	Existence is inherently a synchronistic system.
Linearity dominant.	Nonlinearity and inner space underlying all creation.
Holographic, fractal aspect of universe unrecognised.	Holographic, fractal universes recognised.
Chance, accidental events and luck.	Synchronous events and 'luck'.
Based on the quantitative and separateness factor with only simulated unity; not truly qualitative.	Based also on true unity and the qualitative, and its relation with the quantitative; recognises true unity. Recognises the soul as the next higher aspect (fractal level) of consciousness.

Three-dimensional logic (e.g., whole equals sum of the parts).	Multidimensional logic (e.g., whole greater than the sum of the parts).
Focuses on nonresonant energies.	Focuses on resonant energies.
Recognises emergent software.	Recognises quantum regeneration.
Validates objectivity and intellect.	Includes subjectivity and intuition.
Deals with 'surface' of universe, effects, illusions and linearity.	Embraces the inner structure of the universe and recognises the true nature of nonlinearity.
No science of mind.	Science of mind.
Considers consciousness and mind as by-products of the brain.	Recognises consciousness as primary and the brain (and mind) as a by-product of consciousness.
No life after death of the physical body.	Life after death of the physical body.
Computer system is dimensionally linear (algebraic encoding).	Computer system is dimensionally nonlinear (geometric encoding).
Not recognise relative zero and that the universe functions on a geometry of twelve.	Recognises relative zero and the intrinsic geometry of twelve.

Science laws considered immutable.	Science laws are relative to a limited context and can be bypassed.
The universe likened to a machine made up of particles stuck together by forces—see Figure 1(a).	Universe of interplaying quantum states (wholenesses), in which not only particles but all natural entities have wholeness—see Figure 1(b).
Universe is a closed system of Energy.	Multiverse is perpetual and open; is continuously created.
Static electricity is a closed system.	'Static' electricity is basically dynamic and open.
Learning patterns: spatio-temporal brain patterns and neural networks.	Learning patterns: 4D holographic templates that house programmes and convert nonlinear information into linear information.
Inertia and mass: innate properties; space and time are relative to one another (relativity).	Inertia and mass: contextual; governed by method of interaction; space and time are relative to the many other space-times.
Newton's laws: universal	Newton's laws: limited and can be bypassed.
Forces	Quantum Action
Gravity and the electric field are force-fields.	Gravity and electric field are not force-fields but scalar.

Theories promote ego, competition, and survival of the fittest.	Theory reveals cooperation, support and integration (for proper evolution).
Higher reasoning mind: intellect (no intuition).	Higher reasoning mind: should be a balance between intellect and intuition in complementary relationship.
Communication occurs through 3D space only.	Communication of life and the universe also occurs more fundamentally at a vibrational level (of frequency patterns).
No science of, or a proper recognition of, the species collective.	All species, races, have collectives; a synchronistic civilisation is possible.
Velocity of light limitation.	No velocity limitation.
'Father of Electricity' (false): Edison (rigorously campaigned against AC mains distribution system).	'Father of Electricity' (true): Nikola Tesla (invented AC mains distribution, plus modern electric light, radio (not Marconi), electric motor, turbine, and much more).
Teaches evolution as normal (not recognising presence of a non-harmonic parasitic evolution).	Recognises harmonic civilisations.
Artificial Intelligence. The big (incorrect) question: At what point in the advancing computer/robot system can we *consider* that it is alive?	Artificial Intelligence. The big (corrected) question: At what point can we say that it *is* alive?

Negative resistance, antigravity, perpetual energy sources, cannot be accommodated and understood with the limitation imposed by official scientific laws.	These 'heretical' concepts are a natural consequence of a harmonic science, such as is the New Science.
The scientific observer unknowingly quantum reduces coherent higher orders down to the material everyday level of organisation and randomness.	The New Science recognises the higher orders with appropriate levels of consciousness, *avoiding* the secondary collapses of the wave function (see later sections).

10.

QUANTUM PHYSICS
Clues for the New Science

Quantum physics recognises that all is energy, which has frequency, and at least within the quantum realm that there is no true separateness; all is part of a process—everything is interconnected (see vortex diagram, Figure 4). Relativity, although it deals with a 4D space-time inseparability, no separation between 3D space and the time dimension—viewed in a 4D block form, the ratio of space to time varying according to the observer's motion—it still puts the 'lid' on 3D. This is a little like the analogy of having a see-saw in which the height of the fulcrum determines its dimensional level. Although the ends of the see-saw can oscillate higher than the fulcrum, relativity physics, in effect, fixates the fulcrum precisely at the boundary 3D - 4D. Thus this is not the larger view of 4D, and the special relativity, in establishing the apparent limit and constancy of the velocity of light, simply reinforced the barrier between 3D and 4D, thus still confining science to a 3D world. String theory (dimensions 10 and more) is again a different view of dimensions—minute curvatures at the micro-level, but still representing only an overall dominant 3D existence.

Quantum physics, in revealing a universal and ubiquitous structure of frequency patterns, repeatedly opens the window into the new physics, which is literally the energy window from 3D into higher dimensions and spectra. Even without going into the more metaphysical aspects of quantum physics, the latter still produces the best results with orthodox science. It gives accurate predictions about operations of lasers, microchips, the reactions of chemical elements and the Periodic Table.

At the sub-atomic scale there is only probability, no Newtonian certainty, except statistically in the larger view of everyday life. This gives scope for selection from many possibilities and therefore the potential for choice between the dimensional fractal boundaries.* Clearly these macro-fractal boundaries are created from the whole (Source) and not from the particle, except that the particles comprise the basic units directed by the whole for evolution of universes. [*One can, however, run into certain difficulties when analysing free will—see later section on free will.]

The most important features of quantum physics from the consciousness point of view are the Copenhagen Interpretation, nonlocality, and probability at the micro-level. The Copenhagen Interpretation is particularly interesting as it indicates that the observer, even the observer's consciousness, selects reality from the quantum reality. Interaction itself of consciousness with the quantum realm quantum reduces many possibilities (with different probabilities) to one actual event—which we experience. Even the observation of a particle is a process of selection from maybe countless possibilities from its wave function in the quantum reality. It cannot be assumed that the particle is taking or has taken a particular path before observation. See later section on this.

Nonlocality refers to the instantaneous interconnectivity between all particles that began with a common origin. This has been proved many times but culminating in the very accurate Aspect's experiment. All we're interested in is that this can have agreement with the nature of consciousness—we can see that from Figure 1(b).

We may realise that the Copenhagen Interpretation accommodates potentially the phenomenon of free will, in which the observer selects from the quantum reality of infinite possibilities, bringing about the so-called collapse of the wave function, resulting in the most probable event being materialised. This ties in with the notion that everyone creates their own reality (at some level of consciousness)—see later sections.

There is still, however, within the scientific community, a huge emphasis on classical physics, even if not admitted consciously, which makes it probably more difficult to explain the quantum world to a mind educated in this way than to a beginner.

The classical billiard-ball interpretation of particles and atoms with simple predictability is a very convenient way of looking at the universe. This would probably not be the case if scientists, in general, studied the features of quantum physics. After all it should be clear that all subjects of science become physics as they advance. Imagine biologists, archaeologists, geneticists, or psychologists, thinking in terms of frequency patterns or the interconnectivity of everything in the universe. This automatically expands and channels their consciousness into new avenues of thought, extending the scope when searching for solutions. Even so, classical physics of Newton and Maxwell's electromagnetism were successfully applied to all tested physical phenomena. [But whereas Newton's laws are genuinely limited, Maxwell's original equations, reaching beyond those limits, were curtailed by (mainly) Oliver Heaviside of their scalar, quaternion components, described once as aetheric, which if validated would have linked materialistic physics with the higher spectra, 4D, and Maxwell's equations would have taken us into the New Science.]

Quantum physicist professor Bohm's theories fit well into the new knowledge, with his recognition of the holographic universe. He likens the universe to a giant hologram, in which everything is a constituent component in it. Any portion reflects the whole and vice versa. Everything is an integral aspect of the whole. Bohm recognised the hierarchy of order, what we are calling the fractal aspect of the holographic universe. His 'explicate order' or unfolded order, is perceived by the five physical senses, and his 'implicate order' or enfolded order, corresponds to our inner space dimensions. These theories are required to explain the nonlocality revealed in current science; the instantaneous interconnectivity of events, or what we might call synchronicities.

David Bohm went further and postulated that beyond the implicate order was a super-implicate order. Thus if there is a second order there could be countless implicate orders—as in the New Science, of increasing degrees of order back to Source, the pure 'sentient' state. This is all inherent within the fractal hierarchy. There is always the underlying gradient of increasing degrees of integration back to Source; also providing alternate objective/subjective layers of existence and consciousness manifestation. Furthermore, Professor Bohm's thinking, based on quantum physics notions, arrived at a certain conclusion that all consciousness of mankind was one. That there is no real evidence in the universe of separateness (other than the illusions of relative separateness). This of course has been claimed throughout the history of religions and philosophies of ancient wisdom and, in particular, the present New Age.

Bohm particularly recognised that 'undivided wholes' of energy (quantum states) exist on greater scales than those of atomic physics, such as within planets, stars, or universes. Thus quantum physics supports the New Science, in particular, in this respect.

Why would the old science oppose so vehemently the New Science features? If there is truth in the New Science then mainstream science would surely recognise it?

If current science were truly and solely seeking truth, yes, we would now be educated with the New Science modifications to the old science. Science is heavily influenced and controlled by covert politics, big business, greed and, in particular, the energy crisis. Education and the media, especially television, are controlled sources of information for the masses, including 'excellent' science programmes on the television. Once programmes and belief systems are embedded in the unconscious they can be extremely difficult to erase or realign. The mind can be programmed to believe and do anything. Orthodox science has little understanding of this or how it is achieved. One can do a doctorate in experimental psychology and find there is no teaching on how brain-washing occurs. You won't find the following brief description of this subject in any text book on the planet.

There are two main operations to bring a mind under the total control of its (externally acquired) information, 1) to mould consciousness to the required degree by repetition, hypnosis, or pain subjection and reducing consciousness, and 2) having trapped the mind's focus of thought into its own structures by (1), to make sure it doesn't step above this context by accessing its higher aspects, fractals, of consciousness (for example, soul level); thus a barrier is created between the 3D spectrum and 4D spectrum and above. This already exists to a large degree for the whole population. It results in the lack of awareness at the ego-level of the soul-self (second fractal level).

Let us finalise this section with a brief summary of some striking features that corroborate the connection between quantum physics, the quantum state and quantum reality, with the underlying all pervasive consciousness—that which is eternal and of no location: 1) For a start the quantum reality is an ideal mathematical background for a scientific introduction to the infinite sentience; 2) the statement by some leading quantum physicists that all electrons are identical, in fact, appear to be one electron—one big electron, 3) prior to observation of a particle, it does not have a location, it can be anywhere (or everywhere; remember the sentient reality of no space or time), 4) the term 'holographic', frequently thought to be the nature of the universe and life by many scientists, 5) the philosophical maxim 'everything is everywhere at once' (the sentient reality, which also is the same as infinitely holographic or a perfect unity). All these properties have a relation with nonlocality considered by Schrodinger, 'father' of quantum mechanics, to be the most important feature of quantum physics, 6) the Copenhagen Interpretation and collapse of the wave packet and its relationship to creating one's own reality; see later section.

11.

HARMONIC AND NON-HARMONIC SCIENCES

The need for a harmonic science

If we analyse the knowledge in our two categories in Section 9, we may arrive at the conclusion that the old science relates to non-harmonic science and technology, and the New Science aligns with harmonic science and technology. The latter (harmonic science), is the basis of sacred science or sacred geometry. These have nothing to do with religion, but vice versa religion comes from sacred science (and should be one subject with it).

What does 'harmonic science' (leading to harmonic technologies) mean? It means: no pollution, extremely efficient energy generation, no dangerous energies to life and nature, no radiation, and an abundant and unlimited supply of energy resources. One might see now why the laws of harmonic sciences are suppressed. Most educated people, precisely due to their programming, unfortunately genuinely believe that these systems are impractical. This is totally based on the conviction that mainstream's scientific laws are immutable (constant), which the New Science refutes.

In Figure 5 we see a comparison between a non-harmonic system of propulsion, the rocket, and a harmonic one, the spacecraft. The rocket uses sheer brute force to push against and overcome the inertia and weight of the rocket and impel itself through space, accompanied by various resistances. In contrast the spacecraft propulsion system entrains all atomic oscillations within the craft, including those of the crew, quantum regenerating a whole (vortex) oscillation encompassing the whole

The Emerging New Science

body of the ship. Space is filled with oscillations from not just the craft's vortex but planets and celestial bodies in general. The vortex is adjusted to harmonise with appropriate spatial oscillations. Two or more energy systems become one when made resonant (on the same wavelength), hence the term one-bodied system—the field of the craft harmonises with space (Appendix E). Our technologies for creating force and energy are essentially non-harmonic—the laser is an exception, it utilises coherence of light rays, quantum regenerating enormous power, clean and efficient. More information on technologies later.

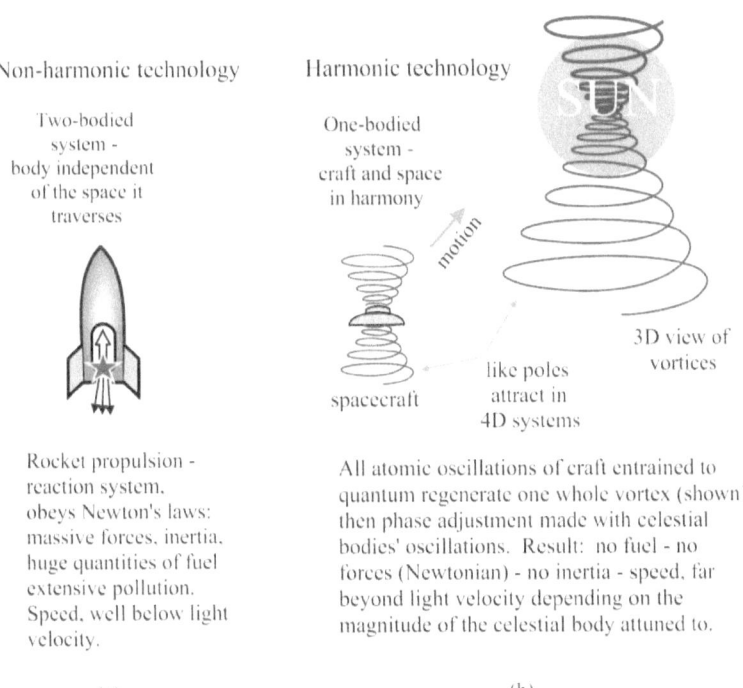

FIGURE 5

12.

THE MIND, CHAKRA SYSTEM AND DNA

Linking the spiritual and the body

The aura layers around the human body, like ovoids within ovoids, are created from the vortex system, bringing in the energy spectra from the immediate upper layers of the dimensional fractal hierarchy. Through these superimposed aura fields the chakras are formed, receiving and transmitting energies, focussing the appropriate frequencies into the regions of the body. The chakras similarly are spinning vortex structures: seven main body chakras and eight morphogenetic.[11] These play a major role in structuring the body and mind. These vortices form links to larger-body vortices and in turn, the planet's and solar system's vortices, and so on.

Most of these energy structures can be considered as part of the mind but as indicated later, the mind, also consisting of memory storage, all recordings, learning patterns, is an extension of pure consciousness (a portion of the non-quantifiable Absolute —see later sections). These extensions are structures of waves, particles, space and time. In the New Science it is recognised that mind/consciousness has higher-fractal aspects (second level is the soul level), extending into higher spectra and dimensions back to Source. Thus one can see that there is always a higher-order level within the human to evaluate any aspect of the universe.

The direction of the primary creation current is 'top down'. It is interesting to note that on each fractal level, which has its

The Emerging New Science

own polarity, the energy currents enter each level below in a neutral state (central) prior to the swing from positive to negative on each level. And vice versa, on the way up (ascension, evolution), the opposite polarities merge in the centre (a more advanced condition), which then moves up to the next level; see later sections.

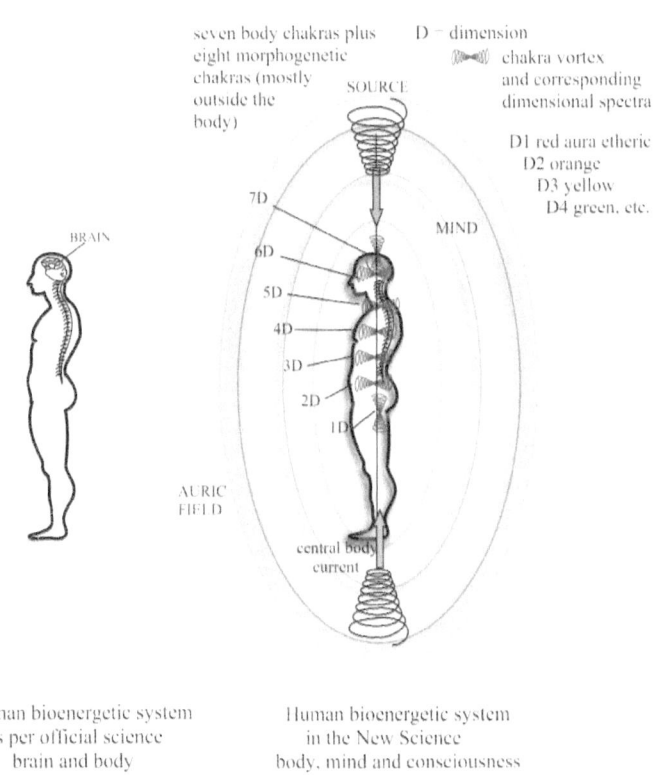

FIGURE 6

Now the chakras also coordinate with the DNA and bring in required frequencies for its evolution; note the dimensional connections between Figures 6 and 7. Also keep in mind that scientific instruments cannot even detect the basic 3D spectra

(D1, D2, D3) let alone D4 upwards. See Part Three, Section 2, and later sections on quantum reduction.

Over 95% of the DNA is deficient of base-pair molecules and these regions are called junk DNA by the scientific community, not realising that the molecules have become disconnected from the higher strands and broken off completely. This is a severe mutation caused by genetic interferences (a major one mentioned in Section 5: *A Glimpse into the New Science*).

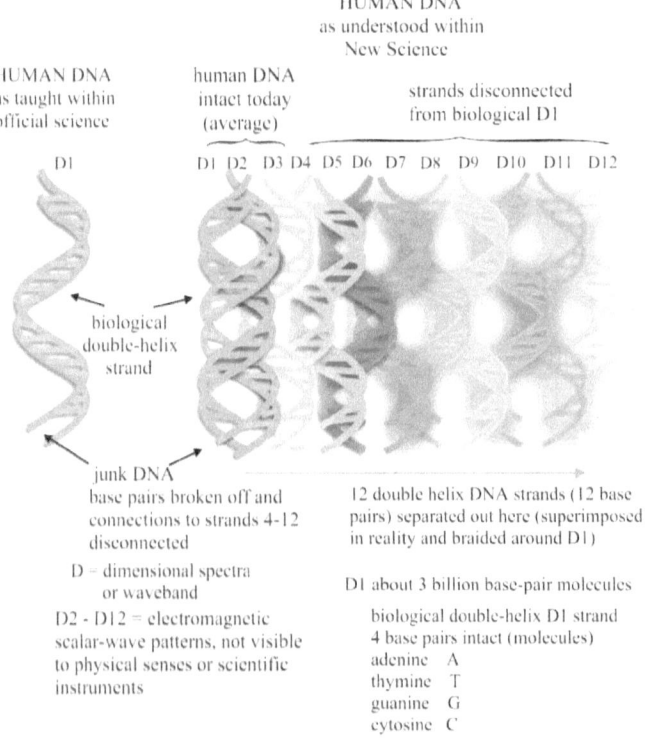

FIGURE 7

The state of the DNA tells us, if we haven't already realised it, that our evolution is not remotely normal, and that in fact we are in an extremely non-harmonic and parasitic evolution (see

Figure 7). Scientists should be asking the question, Is nature natural? Imagine all creatures acting symbiotically—all aiding one another. This is the true path to higher intelligence and proper evolution—which is integration of consciousness (see later related information).

When geneticists discover what the junk DNA is—and it may have already been suspected—they will be ridiculed, attacked and destroyed if necessary, and it will be a much bigger cover-up than the UFO scenario. If the information was released it would change just about everything on this planet—in particular, our history and origin.

Notes

1. www.nhbeyondduality.org.uk. Article: The New Education: *The Paradox of Progress.*

2. Ibid. Article: *Suppression of Knowledge, Discoveries, and the Free Energy Problem.*

3. Book: *Wholeness and the Implicate Order* by David Bohm.

4. Book: *The Original Great Pyramid and Future Science* by N. Huntley.

5. Workshops by A. Deane. The MCEO Freedom Teachings Series, www.azuritepress.com.

6. Book: *Voyagers* vol. 2 by A. Deane.

7. www.nhbeyondduality.org.uk. Article: *Experiential Probability Dynamics and What Happened to Planet X?*

8. Ibid. Article: *There is No Death - You can't Die if You Try.*

9. Ibid. Article: *The Basic Energy Unit: The Vortex.*

10. Book: *The Original Great Pyramid and Future Science*.

11. Book: *Voyagers* vol. 2 by A. Deane.

PART TWO

THE FRACTAL TREE

A Simple Model of All Creation

INTRODUCTION

We say a 'simple model,' which is also an analogy, since the tree is readily understood and can be visualised, and we are using it here to simplify very difficult concepts of nature and the environment. This fractal tree model is intended to assist greatly in understanding the following profound issues: reconciling science and religion; what the soul is; the higher aspects of consciousness of the human; the 'within' (Biblical term used by Jesus Christ); inner space and a fractal system of universes; the true path of evolution; what ascension means; the basis of fractals, and more. We shall see that the tree is more than an analogy but an ideal model for all creation.

It is not necessary to have already an understanding of fractals. The simple tree model will be self explanatory. To begin, we shall consider the tree devoid of leaves for simplicity, and one with a dense structure of branches and twigs.

The first feature we must recognise is that as we go from the twig level towards the trunk, we see that integration within the branch system occurs. For example, maybe three twigs are attached to a branch; similarly for the branches attached to the larger branches as the system integrates and reduces to one item in the centre, the tree trunk. This is a fractal configuration. Let's make the twigs the first fractal level, then the second fractal level is the branch the twigs are attached to, and so on, with fractal levels 3, 4, 5 . . . . For an average tree there are about seven fractal levels. See Figure 8.

Next we imagine that our civilisation is at the evolutionary stage represented by the first fractal level, the twigs. Our full external 3D space, which we identify as our universe, is represented by the twig level only, that is, the arrays of twigs around the tree rep-resent our 3D universe space. Where the twig is attached to a branch, this is the direction of inner space; also the 'within' as referred to by Jesus Christ. Thus we see increasing degrees of integration as we go from many twigs to decreasing numbers of branches to a single source, the tree trunk.

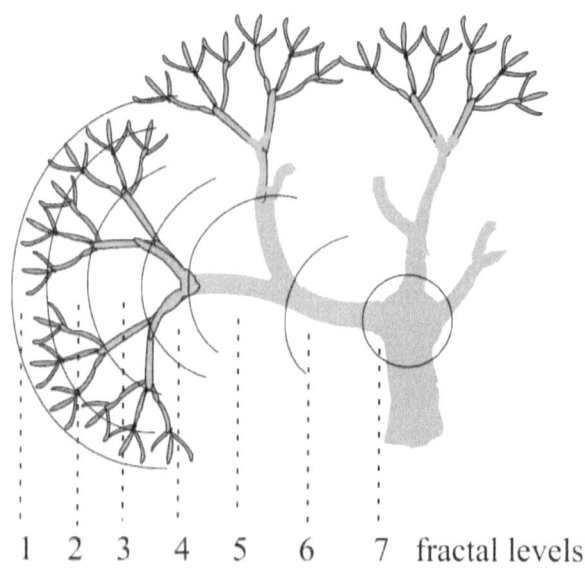

FIGURE 8

13.

SCIENCE AND RELIGION

Science, in viewing the universe (analogy the tree), insists there are only twigs, that is, the tree consists of twigs only. Note that in this analogy, the twig has a frequency spectrum that is the basis of the 3D material world, and this frequency spectrum forms the basic energy and information of this (our) world, just as a waveband creates the television channel picture.

We might wish to imagine, say, an individual who is very short sighted who approaches the tree (no leaves). Owing to the limited perception his vision range only encompasses the twig level. He walks around the tree and determines that the tree consists only of twigs and thinks there is nothing beyond the end of the twig where it attaches to the branch. This point is automatically taken as a zero—there is nothing beyond, and in terms of energies, it is the 'end' of the universe.

This is the 'established' scientific viewpoint today, since it is based on physical senses and scientific instruments, both of which are based on the twig level (but 'dark' matter would be evidence of the higher branches). We shall come back to this.

We may now see that the second fractal level—the branch that the twig is attached to—is the soul level. The soul is no mystery; it is simply a higher aspect of consciousness, existing in the next frequency spectrum or dimension that is higher in rate of vibration than the first fractal, or twig level.

The next branch or third fractal level would be what we might call the over-soul, and so on all the way back to the Source, the tree trunk—the God concept. We now have a comparison of the difference between our observed external world as observed by the physical senses and scientific instruments, and the inner space, or within, as we go from the twig to higher aspects and fractals of consciousness, soul, over soul, etc., to the tree trunk or Source of all energy.

We are now in a position to see the relationship between science and religion. The importance of this is completely underestimated. A civilisation is as retarded as its lack of understanding of the relationship between science and religion, or vice versa, is as advanced as its understanding of the relationship between science and religion. There is really no argument between them; they are each referencing a different context. Science recognises only the twig level of the tree; religion aspires to the higher fractals. Even the conflict between the atheist and the religionist, regarding belief in a god, is fictitious or simply illogical. The atheist is looking for an external God—a very large twig somewhere at the twig (external world) level around the tree. Modern Christianity generally has not realised that this is not their god. Note that to the degree that religionists believe that God is out there, separate, objective in the external world, they will be disempowered. The true nature of the God Source is within—through the fractal inner space system, culminating in the tree trunk or source. Thus, as with the atheist, Christianity supports the notion of an external God.

Once scientists recognise that fractals begin in inner space, nonlinearly, they will have to admit the indirect proof that the fractal system goes back to one source. If there is mere recognition of the system of several twigs attached to a single branch, then several branches attached to a larger branch, it doesn't take a mathematician to realise that the progression is towards a single state at the centre of the tree; the tree trunk. The atheist and the religionist now recognise that they are not talking about the same God. At this point the religionist is quite happy—as long as there is a God, that's all that matters. The atheist, assuming he or she is not dogmatic, will now cease arguing but will next look to see if he or she can believe in an 'internal' Source within.

Atheists are generally intelligent people and will go for science when scientific or logical explanations are available. In this case, if they understand something of the incredible physics of the 'within', in other words, creation, they will go for it.[1]

14.

THE EVOLUTION OF THE SPECIES

The evolution of the species is only a tiny part of the evolution of ascension of consciousness. Why doesn't science recognise an evolution of consciousness in its own right—rather than as a by-product of physical evolution? The reason is that mainstream science does not recognise that consciousness is substantial—a real energy state. It is not insubstantial as is our computer software, which is non physical and is just simply a by-product of a special design of the hardware. What we are saying is nature's systems operate on geometrical intelligence—shape of energy is information. Thus nature's software is also hardware. For example, qualitative states are not intangible, imaginary illusions and by products of that which carries them; they are actual states of energy/frequency, such as for instance a musical tune. Main-stream science essentially calls a melody an illusion and a mere effect or by-product of a sequence of notes. Whereas it is of course the tune itself that is causative and comes first, not the material hardware—the sounds, strings of notes, musical instruments.

Using the tree model, the twig fractal is our 3D level of consciousness. But it has higher aspects of itself, the branches, all the way to the trunk—the source of all the fractal branches and twigs.

The higher energies/information come down from the branches and trunk into the twig, and in fact keep the twig alive; similarly this 3D is constantly sustained from inner space (the inner fractals within). Note that unless the twig is severely mutated it won't 'do' anything to harm any part of the tree or the whole. A mutation would mean some harmonic links have been broken and the twig is not in full communication with the branches, or the links have been bypassed, forming closed loop systems.

In general, consciousness at the twig, first fractal level, can become sufficiently aberrant, misinformed, programmed or arrogant to cause it to consider it is a complete self, and that there is no such thing as guidance from higher aspects (all the way to God Source, the tree trunk). It is now highly egotistical, and is actually in a condition of dying spiritually; being cut off from the source. 'Spiritually' simply refers to the higher intelligence of the branches and trunk. Note that the ego-self is now in a closed loop system with itself (wrapped up in self). We shall see a symbiosis of this ego-self condition with science and technology; they will follow the same principle.

Thus the ego consciousness becomes a closed loop system within 3D. It has severed itself from higher aspects and has no guidance from its own higher intelligence. Picture one twig in this condition acting for self. It has no direct knowledge of the adjacent twigs across space, only objective knowledge—passed across space. It will therefore act selfishly towards them; even harm them. However, if the twig is connected up to the higher branches, as we trace this path from twig to branches we may see that the adjacent twig across space also traces back to the same point. This means the information at this point is a higher aspect for both twigs (or more). That means each twig now has perception of the other and won't harm it. If one goes far enough in this hierarchy towards the tree trunk eventually all twigs come under the same information—see Figure 1. But all these fractal levels are flowing information to the twig level. Thus if the 'twig' is advanced enough spiritually it can sense more branches towards the trunk and finally the whole tree, and would act altruistically for the greater whole.

15.

SCIENCE AND TECHNOLOGY

We have briefly covered the application of the fractal tree model to consciousness. However, there is a symbiosis between science/technology and consciousness. We don't notice it because our conditions on this planet are so much out of balance.

Imagine again our 3D is the twig fractal level. The twig has its own information, but flowing into it are higher levels if information (such as 4D) from the next branch, fractal 2, and from the other branches to the source, the tree trunk. Science only detects the twig information level because of the belief that the only acceptable way of acquiring truth is through scientific methodology. Since scientific methodology or the experimental method is conducted by observation using scientific instruments and physical senses only, then it will only detect the twig information. Physical senses and scientific instruments are in the 3D frequency spectrum range and will not detect the information coming from the higher fractal levels of higher frequencies spectra. Twigness will only detect twigs. In addition, science unwittingly quantum reduces higher states of coherence (and therefore knowledge)—see later sections. As a result of this experimentally imposed limitation, the scientific world goes no further than the end of the twig, a fractal boundary, where it meets the next fractal level, the branch. This boundary will automatically be considered a zero point. Whereas if one were looking from the branch level, outside the context of the twig, one would see that this point of division is not zero. Scientific observation is inside the system of the twig and will therefore only get relative results. This is what quantum physics discovered about 60 years ago and stated the observer is part of the system (same as 'inside' its own context). One must be outside the context of a system to evaluate it properly. These relative results would be the same as using a metre for measurement but having the zero point of the scale shifted to, say, one unit that has been marked zero. Results will be correct relative to one another but all biased by one unit.

If a more consolidated understanding is required, using the same tree analogy, we can imagine one whole branch of this tree to be swaying in the breeze. If one analyses this motion, the twig is executing the greater motion because it is the extreme fractal on the end of the branch. Consider now that science is making a measurement in this 3D universe, the twig level. Since it is inside the context of the twig, where the twig meets the branch will be a 'zero' point—apparently no motion (in the wind). The amplitude of the twig will be measured relative to this zero—like setting the zero on a metre scale, mentioned above, before making a measurement. But we can see that the next fractal level, the branch, is also moving, and carries the twig. Thus the zero point is not stationary—and we have a false reading. But it is acceptable as a first order, until one develops higher knowledge to embrace the branch. Clearly this repeats again and again, all the way to the tree trunk.

The twig fractal is nevertheless receiving information/energy currents from the higher fractals, which keeps it sustained. But science, confined to the twig (3D) level, forms closed loop systems within this level, and does not recognise they are being perpetuated from higher levels. Consider electricity. This is a 2-wire closed loop artificial system. It has to feed off other systems—such as a battery or generator. Inevitably it is a non harmonic technology, since it is divorced from the natural currents of creation, which come down the branches from the tree trunk.

Non-harmonic or harmonic systems refer to technologies for power production, but also the actual structures themselves. An artificial structure such as an automobile and its engine is entirely a closed loop system, requiring energy for power; for example petrol, and force for maintaining the structure: parts held together by forces, nuts and bolts and welding. A harmonic technology is an open loop system tuning into the internal ('within') natural currents of creation, providing perpetual energy. Such a technology may, nevertheless, be combined with non-harmonic technologies based on forces and Newton's Laws[2] (see later sections).

16.

THE INFINITE ENERGY SOURCE

We have already discussed this to some degree—it is of course the subject of free-energy devices, over unity energy supplies, etc., very much suppressed on this planet. In the tree analogy the infinite supply is the energy/information coming from the trunk through the branches system to the 3D twig level and tuning into this. The analogy may be easy to see but what about the physics? Can we present an energy/particle description without getting too technical?

Let us try a couple of analogies. The first one is less efficient than the second, but possibly an easier introduction to the elements of the problem. The components for the non-harmonic technologies are a closed circuit and static field and the requirements of a continuously available artificial input.[3]

Imagine viewing a waterfall from a distance. For the analogy, we just see a column of white water with a constant head (the top of the waterfall column). Consider that we can't see the incoming flow from the top (river), or the exit flow at the bottom. It is just a head of water that appears static. This column of water can be an analogy for, say, a charged battery, in which the two terminals create an electric field—considered static in electrical engineering (ignoring the fact that it is actually dynamic—as we can see in this analogy). However, we know that as we examine the internal structure of the waterfall it is dynamic; there is a continuous flow from top to bottom. Similarly, the battery or any pair of charged terminals (dipole) is drawing in energy from space/vacuum (zero point energy) and releasing it radiantly as observable photons (electromagnetic radiation). The initial energy consists of unobservable particle from the virtual state. This continuous flow within the 'waterfall'/battery from unobservable virtual particles to observable photons is referred to as asymmetric, whereas the static field and the charging and recharging of the battery is referred to as symmetric.

As soon as the battery is charged, this asymmetric flow occurs keeping it in a charged condition. It could be used continuously without depleting the battery—the waterfall column does not need to be destroyed/reduced to provide the energy (flow of water to drive a turbine), there is already an endless stream of water. The continuous flow of water in this analogy corresponds to the continuous supply of energy from the fractal tree source, the trunk, through the branches to the twigs. Whereas the 'static' waterfall corresponds to the energy that only the twig possesses.[4]

Let's now give the second analogy, which may fill in some of the gaps of the first example. In Figure 9 we see a tank of water. This gives us an analogy for voltage pressure since the head of water (subject to gravity) gives a pressure that forces the water through the bottom pipe when the valve is open, and will provide motion for the water wheel or turbine to provide useful power.

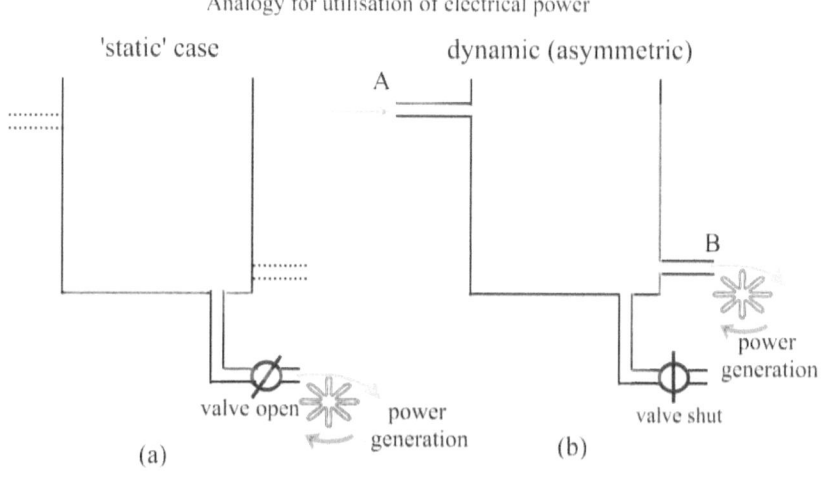

FIGURE 9

Current technologies only recognise the configuration in Figure 9(a). As we use the power from the water pressure the

The Emerging New Science

water drains from the tank, and we have to then refill the tank (which takes a little more energy than we obtained from the release of the water). This is conventional physics converting energy from one form to a useful form but having to input more energy in order to provide a useful continuous supply.

As with the waterfall analogy the tank of water (compare waterfall) corresponds to the considered static electricity—the electric field of the battery and the subsequent voltage pressure. If, however, the internal physics of the battery (or generator) is examined, that is, what is happening within the charged condition of the electric field around the terminals, we would find the dynamic condition of a continuous flow (compare waterfall) through the system. This is illustrated in Figure 9(b). We see that the tank has (all along) two additional pipes A and B, providing a continuous flow of water which is filling the tank at A as fast as it is releasing it at B, achieving a state of equilibrium. Thus the tank is always full once it is filled. Filling it would correspond to charging the battery (it takes energy). Imagine pouring water in the top of the tank to bring the water level up to the top (fully charged) to provide a good pressure as the bottom (voltage pressure). Once it is full it will stay like that indefinitely. Thus again we have the endless dynamic flow from the higher branches of the tree com-pared with the limited energy of the twig.

Unfortunately our current technologies don't recognise this and we open the valve at the bottom to provide the water flow to the water wheel, we lose the head of water and have to refill. Clearly all we need to do is to shut the valve permanently (or not have one in the first place) and place the water wheel at B. The wheel or turbine will provide a continuous source of power since the head of water is maintained.

All analogies are limited. The reader may need to recognise that in the actual energy situation such as in the battery (a process at the atomic level), the ingredients of the water flowing through the tank is different from the ingredients (causing the electronic flow) providing the pressure due to the head of water. This is why the head of water is required in the analogy.

Thus any electrical charged condition creating that apparent static electric field is actually being sustained by a dynamic flow in which the positive and negative poles (kept apart) draw virtual particles from the vacuum and convert them to photon radiation, which radiates out to the universe.

This energy process of virtual particles to observable photon radiation giving a dynamic field and not a static electric field is referred to as asymmetric or broken symmetry, whereas the static field and so called conservation of energy is referred to as symmetric.

This asymmetry was discovered in 1957 by Lee, Chang and Wu and a Nobel Prize awarded. In spite of that it is not included in electrical engineering and all energy sources must conform to the symmetric.

The Maxwell's equations taught in universities and, in particular, to electrical engineers are not the original Maxwell's equations, in which the asymmetric terms were removed by physicists Heaviside and Lorentz, and a more advanced physics, leading to a harmonic science with free-energy sources was buried.

Thus there is the need for recognition of higher energy spectra—the higher fractals in the tree analogy and the realisation this is a continuous supply.

17.

EGO/GOD DICHOTOMY

We have explained that the source or God is not external but 'within'. In fact, if we are at the twig level and God is the tree trunk (but also its extensions) then this means man's highest aspect or fractal level (twigs to branches to trunk) is God; also, that all creation, whether an atom or sentient being, is an extension from that source.

What is the ego? The ego should be merely the identity at the twig level focussed mainly into the twig fractal level, with natural universe boundaries between 3D and 4D or above, that is, between the twig and the branch to which it is attached. The higher levels would nevertheless be available to the ascending 'twig'. However, with all the aberration, mental illness and mutations on this planet the interconnectivity between the twig and the branch is particularly blocked. The 'twig' identity consequently becomes isolated from the tree/universe, and subsequently operates on, and emphasises, self—since it does not have perception of and from its higher fractal levels. The ego becomes a closed system and automatically will create closed systems (non harmonic) technologies (symbiosis between consciousness and technology).

Let's take religion, in particular, Christianity, although it has no knowledge of the way these energies work it nevertheless teaches the opposite of ego behaviour. The statements like: 'Doing the will of God,' or 'Let God take over your life,' have an authentic physics explanation—it is simply illustrated in the tree model. These phrases that appear to many people as though one is giving up self, which feels wrong to them, simply means endeavouring to recognise that one's higher aspects of consciousness are connected and part of the Source/God/tree trunk and to tune into these energies and probabilities—the creation currents from source.

This is what the true self is all about. Instead, the ego, focussing too much on self, fails to detect the continuous guidance from the soul level (branch) and higher levels, and forms closed loop systems, causing the ego to be extremely vulnerable to negative influences: making errors, going astray, committing crimes, etc., mainly because it can't take into account the bigger picture and therefore consequences of actions.

The ego as the twig will only perceive another person (twig) across space (for example, one twig next to another). It can harm or be selfish to that person since it does not sense their viewpoint or how they feel. If, however, one is in tune with the higher aspects, for example, the branch to which the twig is attached, this has a common point for both twigs. The person is experiencing energies/ information from a level now that is the 'collective' or unity of both these people, or twigs. They are sharing the same aspect/region of consciousness. If one is tuned to this higher fractal, which also means is receiving at the twig level the information from this highest point (higher frequency), one will have part of one's consciousness in common (even if temporarily) with that person, and couldn't harm or make (direct) mistakes which would be a disadvantage to them. Thus the branch level is inherently there at all times and could be activated by appropriate harmonic behaviour at the twig level. Think of these energies as frequency patterns.

The ego (the twig level), if sufficiently strong, feels uncomfortable with the idea of having higher aspects—the upper fractal levels. It feels threatened since it can be controlled by these higher levels. But these higher aspects of consciousness would only have the best interests of the 'twig' at heart. However, these interests, goals, purposes, programmes are not materialistic primarily, which is all the ego is interested in.

Thus the ego will oppose the higher information and opportunities, cutting itself off more, but also will attract (be vulnerable to) negative manipulations that will sever it still further (creating a block between the 3rd and 4th dimensional spectra).[5]

18.

CONSCIOUSNESS AND THE PARANORMAL

The twig level analogy has a spectrum of frequencies; this would be in the 3D range. Paranormal phenomena, ESP, clairvoyance, come from the higher-frequency spectra above the 3D level, which we can call higher-sensory perception. This would correspond to the second fractal level or branch above 3D or the twig. Clearly if physical senses and scientific instruments are structured out of the frequencies or lower wavelengths of the twig level they will not detect the branch level or higher fractals.[6]

The higher fractal levels nevertheless interpenetrate down to the twig level but not everyone can detect them—and of course science denies them since it can't detect them with the limitations of physical senses and scientific instruments. Those people who can detect those higher frequencies already have them accessible in their consciousness spectrum. Individuals have their own unique spectrum frequency distribution—this is their true identity signature. It is a complex waveform of superimposed sine waves of countless different frequencies with different degrees of integration or fragmentation (out of phase energies).

Some people will have a low or high distribution, or be in the middle range; others will have some high or very high components mixed with some low ones. These could be viewed on an advanced scalar electromagnetic oscilloscope (of the future). One would see a complex waveform, maybe in several parts, of various frequencies. An average frequency rate could be taken indicating the evolutionary level of that personality, or more accurately the complex waveform could be Fourier analysed to break it down into individual sine waves to reveal the harmonic or non-harmonic relationship between the sine waves (alignment and phase state).

In the higher-frequency bracket, with harmonic relationships between the waves, would be behaviour states of altruism, benevolence, ethics integrity, appreciation of great art and music. The lowest regions would contain frequencies out of phase—the twig out of alignment with the higher branches. Here we would have selfish, aggressive traits and criminal behaviour. However, it is important to realise that a person may have higher frequencies expressing the above greater qualities, even be a devout Christian, but if they are dogmatic, with rigid patterns of thinking, are intolerant and overcritical of other viewpoints, this will also produce a lower range of vibrations.

19.

ASCENSION

So what does this mean as far as the person's future is concerned, in particular, the next life? Firstly, the higher aspects of consciousness already exist—the upper fractal branch system of the tree—which is somewhat like the future of the lower 3D aspect, the human personality. An analogy for the way this occurs is to imagine white (ordinary) light to be passed through a prism. It is split up into a spectrum of frequencies ranging from blue to red—this is how the rainbow manifests.

The original consciousness of the individual similarly divides dimensionally, projecting a series of fractal levels of consciousness throughout the universal system ranging from a higher frequency wave band to low with a corresponding entity formed at these different dimensions. For example, in the analogy, the lowest frequency band is the 3D, or twig level, the next, the branch to which the twig is attached, the soul level, and then the next larger branch, the over soul and so on.

During ascension, or true evolution, the lowest extension (the twig level) is expected to draw in (attract) the higher frequencies coming through from the higher fractal branches, assimilate these into the personality (individual's frequency spectrum) and into the DNA (connecting bit by bit the biological double strand with higher dimensional scalar electromagnetic strands braided around the physical/biological strands—and restoring the junk DNA to naturally occurring base pair molecules).

One might imagine now the twig being drawn into the branch and absorbed by it, as the twig now manifests a higher state of evolution. The analogy may be observed to reveal that more than one twig merges into the branch—this is not an inadequacy of the analogy but we will not pursue this here, as it is something that the lower ego can never appreciate and feels uncomfortable with. It is briefly covered later. Nevertheless the lower parts of the self merge into the higher parts (fractals) and ultimately go back to source.

The reader may wish to recall now the tree analogy for the big picture of the perpetual energy. The continuous supply of energy is that coming down from the trunk to the twigs, and we might wish to then imagine that the energy, in addition to forming stable structures of energy it also radiates out from the twigs and goes back to the source, the tree trunk, whereas the orthodox view would be having a closed loop system at the twig level. We might wish to imagine for this, a twig, say, kept in a bent position, like a spring, now containing stored, or potential energy, to be utilised for doing work and subsequently depleted with no continuous supply.

People are educated to believe there is only 3D, that is, the twig fractal level; and what one believes on mass will be created—meaning the higher levels will be blocked off. Thus using the above analogy when the stored energy in the bent twig is released, the twig has to be worked on to bend it again and provide stored energy, when in fact its very creation (of the twig) involves continuous energy coming from the higher levels, and ultimately the trunk, or Source.

The answers and solutions to everything are thus in the inner space—the fractal system from twigs to trunk, not the external layer of twigs (our 3D universe). In religious terms, this inner space is the within taught by Jesus Christ. All answers are therefore 'within' and not outside the self; where 'self' here would be in the larger sense going back to the tree trunk.

Notes
1. Articles : *The Fractal Matrix*. www.nhbeyondduality.org.uk.
2. Article: New Education: *Reconciling Science and Religion*. www.nhbeyondduality.org.uk.
3. www.nhbeyondduality.org.uk. Article: *Suppression of knowledge, Discoveries, and the Free energy Problem*.
4. Bearden, T., www.cheniere.org.
5. www.nhbeyondduality.org.uk. Article: *Evolving Civilisations*, Part I & II.
6. Ibid. Article: *Source of Fractals*. www.nhbeyondduality.org.uk.

PART THREE

EDUCATIONAL FALSEHOODS

1. There is no life after the death of the physical body.

2. Scientific methodology is the only acceptable method of acquiring truth.

3. The origin of the universe: Big Bang theory.

4. The origin of the species: theory of evolution.

5. The mind is a by-product of the brain.

INTRODUCTION

It may appear a rather harsh and exaggerated accusation to make to regard these well-researched bodies of knowledge as 'falsehoods', and it is hoped (likely in vain) that the academics against this viewpoint don't take it too personally—there is a degree of truth in the parts—even when not taking into account the whole. Science is generally correct with its carefully and rigorously controlled experiments and measurements, which usually will repeat with consistency and can be corroborated, but we are merely trying to encourage the scientist to look at a much bigger picture—a far more elaborate and sophisticated paradigm that will, when studied and understood, bring together in harmony all conflicts and resolve all paradoxes and enigmas of science and life.

At this level of difficulty in evaluating such sweeping universal concepts isn't it a matter of opinion, one might ask? What about the New Science, couldn't this be based on falsehoods? We can only leave that to the students to decide for themselves. When dealing with higher knowledge one must acquire one's own subjective understanding. Objective scientific experiments will not provide proof. Advanced civilisations (not us) will have trained and evolved their minds to understand potentially anything without the dependency on scientific methodology that we use.

Nevertheless, note that we do not designate a subject or body of information as a falsehood if it teaches it as theory, such as was the case with Darwin's theory of evolution and the Big Bang theory. Unfortunately today these are being promulgated as established truth, in particular, they are expounded on the ultimate mind-control machine, the television (equally it could be a blessing). By repeating something many times the subconscious actually begins to believe it is true, giving an irrefutable conviction to the vulnerable conscious mind. The public should be outraged that these 'truths' are being embedded into the minds of the young. Science, with great altitude and conceit propounds the most 'de-graded' (lowest grade) solutions to these extremely profound and complex questions. However, there is no criticism of these theories as long as they are referenced as theories (and not yet facts), but as indicated above, they are today being disseminated and taught as truths to the uninformed public, in particular, the youth—the new generation. This is a massive responsibility.

How can scientists with no background in mind, consciousness, and spiritual studies evaluate in a balanced manner such vast phenomena as a theory of evolution, Big Bang, and continuous life? To believe that life is separate from the universe is a huge and unwarranted assumption.

Many conflicting theories usually all have some truth. Science generally makes correct measurements; it is the interpretation and the context that is the problem. Scientific instruments have low resolution (for detecting wavelengths) but high sensitivity to intensity. This means they can't remotely detect the underlying higher frequencies of mind and consciousness.

It is time that the portion of the population in the world, in fact, a dominant percentage, who recognise these untruths, made a stand against these outrageous beliefs being blatantly and subtly programmed into the mass mind through education, orthodox scientific sources, and the media. At least if there were a scientific theory of creation it would compensate somewhat for the obvious current biased direction of thought.

20.

THERE IS NO LIFE AFTER THE DEATH OF THE PHYSICAL BODY

Consciousness is an illusion or a by-product of the brain.

The title means of course that we (or other life forms), at the death of the physical body and brain, no longer exist in any form whatsoever. Interestingly this directly relates to whether one has a mind (an energy structure) that is not a by-product of the brain. If it were a by-product, as science teaches, it clearly would also disappear (and that is partly why the mind is being taught as a by-product of the brain).

In the New Science, the mind is a separate entity composed of energy/frequency patterns, and since energy does not die, it therefore lives on potentially indefinitely. Frequency patterns don't require feeding and don't wear out. Continuous existence beyond physical body death is elementary physics and is an inherent and intrinsic part of evolution and ascension.

Throughout history to the present day, one couldn't find more evidence for the continuation of life. However, proof through the experimental method will never be achieved with the current attitude of sheer disbelief without researching the subject.

Clearly this particular falsehood links with the one that implies we are only a brain and a body—Section 5. Surprisingly though both these have a relation to the Big Bang theory falsehood as we shall see. The connection is to do with the big question of whether we can arrive at or find 'nothingness'. Media academics believe the universe began with nothing. We shall see that there is no such thing as nothing.

Relevant to this falsehood, regarding no life after physical death, so many factors of existence provide evidence and support continuous life; there is no religion without it, no New-Age information, and the whole field of spirituality is based within it.

In the therapeutic field there are utterly countless examples of people recalling past lives. Many native populations such as American Indians recall and accept past lives as a norm, including recollections of major events such as the destruction of Atlantis (this degree of memory would be normal if the DNA was more complete—no 'junk' DNA).

Extensive research has been done in past-life regression therapies, including scientific testing of case histories. In addition, there are hundreds of thousands of cases of past-life recall, which occur spontaneously as a result of therapeutic processing within the philosophy of scientology.

Even leading physicists have verified life after death from their observations. There are thousands of examples of 'near death' experiences in which a momentary death has actually occurred and the person could relate the after-life event, after returning to Earth. Dr. Jeffrey Long, director of the Near Death Experience Research Foundation considers that the near-death experience provides the strongest evidence for life after death of the physical body.

Investigator, Michael Roll, has dedicated himself to bringing truth to the public regarding continued life after the death of the physical body.[1] He reminds us that physicists, such as Sir William Crookes, Sir Oliver Lodge and John Logie Baird witnessed repeatable experiments under laboratory conditions in which people who once lived on Earth came back either through voice or astral appearance and proved that they had survived physical body death. The public have no awareness that there is a powerful campaign against providing them with proof of continuous life (which goes even beyond covert politics).

Life after physical-body death is an inherent and intrinsic feature of the New Science. Note that, as already mentioned, how this relates to the later brain-body falsehood debate that if the mind is not a by-product of the brain, to which we subscribe, then life must be continuous. The mind is an energy structure and if it is not a by-product of the brain, that is, independent, it doesn't die.

One of the big questions in science has been whether it is possible to achieve nothingness. Scientists became aware of this quest when fortuitously causing reduced air pressure in vessels

The Emerging New Science

and tubes. Even when the withdrawal of air was accomplished to a high level of vacuum there was then the question of whether a perfect vacuum was empty. We now know within orthodox science that space is teaming with activity—in particular, with minute particles and antiparticles coming and going. They are called virtual particles and manifest as minute wormholes—mini-black-holes going inwards and mini-whiteholes coming out. One of each of these, fleetingly coming into existence of the same size and colliding, will cancel one another and return to a state of 'nothingness' (a zero). The New Science disagrees as we shall see. In the Big-Bang section many mainstream scientists believe these particles and antiparticles come from nothing (one should be surprised that science could entertain something so magical).

What we shall do here is to carry out a thought experiment to illustrate theoretically the step-by-step removal of all mass, energy, particles, waves from a given system to see if there is anything left. We can begin with a living human. The body eventually dies, disintegrates, along with the brain, and disappears. However, we can know that the mind still remains, which is not a by-product of the brain but an independent, far superior computer system—see later sections.

The mind is made up of energy patterns and energy doesn't die, as already mentioned—it does not wear out; it does not require feeding. This should be elementary physics today. However, what we are interested in, is would there be anything left if the mind was wiped out? In theory anything that is quantitative, quantifiable (meaning can be analysed), could be cancelled out or cut off at its source—anything structured and contains space and time.

Thus we now imagine erasing the mind. This would eventually remove all memories, personality traits and identity frame-works. However, it would be a complex process, since we would find there is an input to the mind. What we call a consciousness. This has a frequency structure in itself, as does the mind, and interfaces holistically with the mind, making them virtually indistinguishable. But strictly they are function (consciousness) and structure (mind). Nevertheless this input consciousness has higher (frequency) aspects—greater fractal levels—and higher-order recordings and memories. Thus we now have to imagine cleaning off all these quantifiable features: particles, waves, frequency patterns. Eventually as we erase everything that is quantitative and tangible, these particles, waves and space-time will disappear. Surely there is nothing left now. If there were nothing left, then how did the quantitative aspects of creation, in other words, manifestation itself get there?

We shall find that at the end of this long line of 'removing' finer and finer structures, that are all potentially quantifiable, there is no 'nothing'—there is no such thing as nothing. This is now a state beyond space or time, which means there is no location—just a 'wholeness' 'everywhere' (remember the axiom: 'Everything is everywhere at once'). There is no time—just an eternal condition—simply a 'thereness' with no beginnings or endings. This is the experiential condition of life and existence, and is present within us at all times. Compare this condition with a computer, which is totally quantitative and quantifiable; it does not experience. (Nevertheless, the New Science can answer another big question of science implied here, whether an advanced computer or robot could ever be alive—see computer section.)

Thus we have arrived at the raw, basic, native state of unadulterated awareness, or sentience, prior to all of creation. It is totally alive. If the human had gone through all the erasure speculated above, he or she would have no objective memories, no identity or personality, but would be totally alive, and automatically merged with this basic eternal state of creation of no location, manifesting as one whole. We have thus arrived at a feature of the Source, God Source, a single state, a 'self' from which springs all creation.

Following through on this, by focussing away from the wholeness, thought projects a concentration of energy/particles and the first signs of what is quantifiable appears—what comprises existence with the necessary separation factor and therefore space and time. These same energies and particles condense into 'moulds', or templates, which are frameworks for shaping all forms and structures. This same energy, when drawn into the templates, will accumulate and shape that appropriate form, particles, species, aspects of the universe, according to blueprints. Thus we have a basic blueprint for evolution of existence and the original self-exploration of its own infinite beingness, manifesting through us and all life its infinite probabilities, and forever becoming.

However, we are only interested in identifying this sentience to show that we simply can't die even if we tried, but all structures (even memories) could theoretically be erased, and that consciousness would then be reused, programmed.[2]

This sentience, which we generally call consciousness, exists in all things. A good analogy is to imagine a flexible fluidic medium, such as water, being poured into grooves or moulds (templates) cut out in a board. The grooves represent programming, learning, automatic responses, even pain structures, which imprint or shape the water, thus manipulating the sentience in different ways. All creation operates on templates that shape consciousness energies. Shape of energy is information, nature is not an algebraic abstract system that our intellects are trained for and tend to prefer. We see that this is a form of geometric intelligence.

Life after death has nothing directly to do with religion and specifically 'Heaven', except that you would expect this New Science material to be received agreeably by the religious community. It doesn't mean there is no place/planet worthy of the label 'Heaven' (there is in fact a particular planet in the second fractal level of the universe which has had this role). Unfortunately many religious leaders consider that their positions of control and occupation would be jeopardized by finding that the after-life is part of the science of true evolution, and will attack anything that threatens the exclusive monopoly over their religious followers.

Serious scientific investigators in this field have pointed out that in fact the after-life is a branch of subatomic physics and is natural and nothing to do with the supernatural, religious beliefs or parapsychology. 'Subatomic' at our present level of science is as far as physics goes but the prefix 'sub' can take us further and deeper into the inner workings of creation—the inner space through the aetheric and into the astral and beyond; the domains of other realms.

Amazing how 'cleverly' the aether was removed by relativity. In doing so, the link to the higher-world fractal level—the higher spectrum—was blocked to further scientific developments and to anyone who believed in these higher existences, once and for all. This is like disconnecting the twig from the first branch, which then makes sure the reality of our 3D is isolated. The concept is as simple and obvious as that.

We have a science trapped in materialism (3D) and Western religions (their authorities) who covet their monopoly on the life and death industry, making a further contribution to retarding our civilisation's proper evolution.

21.

SCIENTIFIC METHODOLOGY IS THE ONLY ACCEPTABLE METHOD OF ACQUIRING TRUTH

A fundamental set-back to the acquisition of knowledge.

We can begin by showing that scientific methodology or the experimental method, which means proving theories by rigorously conducted experiments, has severe limitations that are not being taken into account. However, in doing this we won't have actually shown that, no matter how limited science is, it is not the best method. This is another problem. Nevertheless, to know how science can go astray is a great step forward.[3]

We see from Figure 1 that science completes itself with one level of creation (3D), the lowest spectrum in Figure 1(a). This means life, people, the human observational viewpoint or the observing scientific experimenter is also made up of this same frequency or wave band (even referred to in quantum mechanics as a participator in the process of creation). A proper evaluation of a system cannot be made while inside that system (on the same level, which means within that context). A machine can't evaluate another like machine. This is what puzzled Darwin when he stated in a letter (read out on a radio programme), 'How can man judge (understand) nature if man is part of nature?' His question is based on the assumption that existence is made up as shown in Figure 1(a)—man being part of the same substance as that which he is observing.

A very good analogy for this is the well-known saying, 'Can't see the woods for the trees.' One has the viewpoint inside the woods, as do any of the trees, and can't see the boundary of the woods or the relationship of the woods to other woodlands.

One must be outside the context to evaluate it (for example, above the woods). Darwin didn't, and scientists in general don't, realise that the mind extends higher—into higher frequencies and can thus always (potentially) attain an observational state higher than nature or the universe, as we might see from Figure 1(b).

Consciousness and mind extend with their higher aspects up the frequency spectrum and thus can evaluate any context by going into a higher viewpoint. This viewpoint, or point of observation, is governed by the resolution or wavelengths (a property of higher frequencies). The mind is always capable of going higher in frequencies (higher resolution, shorter wavelengths) than the universe region being observed.

The New Science handles this adequately but does quantum physics help here? Indeed, it is in fact one of the main sources. Quantum physics exposed this experimentally some 50 years ago and drew the same or similar conclusion as Darwin's. In this case, it was that the observer (experimenter) was part of the experimental system (compare 'part of nature'). Note physicist John Wheeler's statement for the experimental method or any observation, 'The man who stands safely behind the thick glass wall and watches what goes on without taking part . . . can't be done'. This means the observer is not truly objective in our scientific methodology—thus the basis of the acquisition of reliable knowledge in science is flawed. Why is this? It is because science only uses physical senses and scientific instruments, which are structured within the same spectrum of frequencies as the apparatus (experimental environment, nature, and the universe).

It may be of interest here to mention that invisibility, or the undetectability of energies or objects, occurs when the energy is of a higher frequency, and also when it is oriented differently, not 3D-wise, but in a 4D direction. The military 'Stealth' invisibility goes beyond the screening of radar (not publicised). Nevertheless the invisibility is created by electromagnetic force-fields, which bend light (around the object)—it is still there though and detectable by more direct means, such as physical contact. This was the basis of the Philadelphia experiment in 1943, but in fact surpassed the electromagnetic invisibility (due to intervention from other factions).[4]

In general, however, where the scientific method is applied to observations and experiment, which are relative to a known context, such as measuring the acceleration due to gravity, there is no problem; results are relative to a known context and are correct. But where science is pushing the boundaries into new territories it will only give relative results, for example, physical constants are only relative and will change with expanded contexts of knowledge. This scientific limitation will inherently restrict knowledge and subsequently adversely affect our evolution. Anything will self-prove relative to its own context.

The answer is not simple. Scientific methodology should continue to be used as much as possible but keeping in mind the limited context. The educational system should emphasise more right-brain abilities, intuition, imagination, inspiration—and there should be acceptance of a mode of acquiring information from many sources. Eventually with a powerful development of the intuition, research will move in the most advantageous direction and experimentally untested theories could be evaluated for truth content by the intuition.

In a nutshell, an advanced civilisation has no need for the experimental method; they can use their minds to acquire scientific data. Everything can ultimately be known through the mind and consciousness development, achieved by consciousness 'projecting' into that which is to be understood, since everything is made up of units of consciousness. See Part Four.

22.

THE ORIGIN OF THE UNIVERSE
BIG BANG THEORY

A higher order from a lower order or randomness?

The Big Bang theory superceded the Steady State theory and became the most favoured hypothesis for describing the origin of the universe. This theory proposes that the universe was originally in an extremely hot and dense condition, which then expanded, creating a mass of superheated subatomic particles, and continues to expand to this day.

One of the principal observations which led to the Big Bang theory was that made by Edwin Hubble who discovered that the distances between galaxies were generally proportional to their red shifts, indicating expansion. However, there are other possible causes of the red shift. This expansion explanation is an example of interpretations of scientific data. The observed data is correct but the interpretation may be incorrect. Interpretations involve assumptions that get forgotten about, and science finishes up with a belief system without it being realised.

The red shift can also arise from other conditions. In the New Science, the vortex is the basic energy system underlying all creation—compare a whirlpool on water. The light wave coming into the vortex of the Earth, as seen by an observer on Earth, will be accelerating into the vortex, causing a lengthening of the wave and the subsequent red shift.

The other major feature corroborating the Big Bang is the cosmic microwave background observation, evidencing that the origin was extremely hot. The after-effects of this can be measured today and this is considered to be a remnant of the Big

Bang expansion. This and the red shift are the two main pieces of evidence.

Nevertheless, on the down side, the cosmic microwave radiation indicates that different regions of the universe, widely separated, are too far apart to have come from a common source. The cosmic background theory does not handle this. This is referred to as the 'horizon problem'. There is also the 'flatness problem,' indicating that the lack of curvature of our universe suggests that a Big Bang expansion would not explain this degree of curvature. Moreover, the Big Bang expansion should be uniform and thus does not provide a basis for the presence of the so-called 'clumpiness' of our galaxies. Finally, in order to explain how nuclei were formed in the Big-Bang high temperatures, it is necessary to utilise the Grand Unified Theories, in which instantly after the beginning, the strong, weak and electromagnetic forces were in a unified state, but such a state of indistinguishability of these forces is not found.

Today's estimates of when this occurred are given as 13.7 billion years ago. The Big Bang was not understood to be an explosion but its beginning must have developed from an extremely hot and infinitely dense state, explaining the expansion, including that of space itself. However, the Big Bang as such does not provide any explanation for the initial conditions—just the general evolution after the beginning. There was nothing to explain how it started and what preceded it. The origin was considered as concentrated into an infinitesimally small point of infinite density, which is impossible with current science and can only be an assumption. The very prediction of such a singularity indicates that the theory has broken down (for science's narrow framework).

In the New Science the vortex interconnects the dimensional layers, and thus it in fact shares properties with the singularity. The singularity, to escape its infinitely dense impossibility could possess a 'portal' at its centre as does the vortex. It follows then that all the mini-black and mini-white holes of vacuum (the particles) are minute vortices that bring in energy into this 3D system or take out energy to another. Clearly the conservation of energy requires a higher order that this New Science paradigm provides.

The two dominant flaws, which are all-encompassing, are firstly that the amazing design and order of all creation came from a random condition, and secondly that there is no rational explanation for the initial conditions. Nevertheless a number of scientists, regarding the second point, seem satisfied that the beginning can develop from true nothingness. Clearly the Big Bang theory was becoming as fantastic as any supernatural theory or improbable religious claim. But scientists still had hopes for at least a logical explanation for the existence of something from nothing.

Discoveries in astronomy and physics have continued to indicate that the universe did have a beginning, but leaving the mystery of what came before. One moment there was nothing —then suddenly there was something. Attempts have been made to provide an explanation for this and that it began with a singularity of infinitesimal size, and of infinite density. This is akin to the centre of a black-hole, with its huge gravitational pressure and where the normal laws of physics go awry. But scientists did recognise (and some do quite correctly) the supernatural element of how it all began and what came before—even the question of there being a creator. Desperate for some random materialistic description for this beginning, additional supporting models were sought.

The Inflationary Model is one such model. It does not, however, theorise the initial Big Bang requirement. The expansion can arise from fluctuations in the vacuum, resulting in rapid expansion, occurring in a very short time. Developments in particle physics and quantum mechanical interpretations of the vacuum effects have supported the Inflationary Model, which tells us that in utilising some of the features of the application of quantum physics to the vacuum, particles and antiparticles fleetingly and continuously appear, then annihilate one another (being matter and anti-matter) and return to nothing. But available evidence in physics for these fluctuations does not explain the asymmetry in a universe such as ours, which is lacking antiparticles compared with particles. This is easily handled with the vortex model since the opposite pole of this dual vortex system is the anti-side (anti-particle state to all creation's entities—particles, planets, universes, etc.). We are in the particle side, meaning one side of this oscillation. Particles and antiparticles are balanced in the big picture.

Thus the Inflationary Model, utilising quantum physics of the vacuum and the fleeting appearance and disappearance of virtual particles, takes the Big Bang to the extreme of a universe that evolved from nothing. This apparent violation of the conservation of energy is acceptable in quantum physics as long as the time intervals for the fluctuations are short enough.

As we saw in the section on continuous life there is no such thing as nothing. It can't exist. Scientists are assuming there is nothing beyond the virtual particles that spring into existence, because they can't detect anything. The big shock to science will be when the true role is discovered of the particle and antiparticle interaction. In the New Science, the particle and antiparticle are the fundamental particles of evolution and do not generally annihilate one another. In the laboratory they may cancel completely, as deduced by science, and also cancel due to flaws in nature, caused by malfunctions and mutations in life processes.

The true evolutionary characteristics of these particles will not always be evident on this planet (or local regions of the galaxy). In the proper evolution of nature the particle and antiparticle merge to form a new particle of higher order, meaning of higher frequency and occupying a position in the next higher-frequency band—the next fractal level. On this higher-order status the new particle then immediately undergoes fission to take on the particle and antiparticle role again but with a more integrated relationship.

This continues up through the fractal system (compare twig evolving into branch, into larger branches) to still higher orders of greater integration until it has reached the source of no separation, no space and time as described in the section on continuous life. The purpose of this is to create a gradient of many steps (fractals) from total unity at the 'top' to maximum separation into parts at the bottom. The steps or levels provide different orders of dimensions for the exploration of consciousness, with different ratios of the factors objective/subjective.

Once the scale of this is recognised, and that the universe is a holographic fractal system, then the debates about whether this or that Big Bang version is more correct become superfluous. Further, the anthropic principle also becomes superfluous, or not questionable. That is, when we realise that the Big Bang's limited scientific basis must be replaced by a much larger multidimensional structure of holographic and fractal interconnecting universe layers, precisely as with the fractal branches of a tree, a whole new basis is required in the explanation of creation. Maybe a type of Big Bang phenomenon occurred which in our fractal model would be similar to a sudden initial appearance of the twig level extending from its branch connection. This could in fact be produced by a type of singularity, which connected through the dimensional layers, but they would be white-hole singularities. The infinities of science, such as infinite density, are an illusion due to trying to make such a region a zero point of existence—when it does in fact contain infinity inside and beyond it. A civilisation structured out of this spectrum (3D, the 'twig' level) would only detect and know about this environment, totally unaware with current scientific instruments (of relatively low resolution) of even the next fractal level (the branch).

If science applied universal fractal theory, which would make the assumption that all facets of existence follow the fractal law—already accepted by most scientists who truly understand fractals—the Big Bang theory or any truth in it would be a small part in a greater whole within the multiverse.

The Big Bang theories are riddled with interpretations, even metaphysical and certainly philosophical, with no directly detectable evidence that what science is discovering and measuring is really related to the fundamental features of original creation. Nevertheless, in contradistinction mainstream science will promptly discredit unconventional theories on the basis that they don't have concrete proof. Modern science does not remotely have a sufficient paradigm to be able to draw any commensurate and rational conclusions about the origins of the universe. It simply doesn't have enough relevant technical material, and we are left with having to accuse the human race of sheer arrogance to consider that we are so advanced.

String theories propose that all particles are represented as strings of extremely minute length, oscillating at various frequencies, and requires eleven dimensions—more recently new string theories have extended that to twenty dimensions (above four dimensions these are merely on a micro scale). However, String theory, although showing promise in explaining some major problems of cosmology, in particular, after the Big Bang, it presently does not add much support to the Big Bang weaknesses regarding the prior condition of the expansion and the evolution of order from disorder.

As we may have noted from the fractal-tree analogy, claiming that the universe began with the Big Bang is commensurate with stating that the twig of a tree came into existence from nothing. In this analogy, the branch and trunk that originated the twig are in the inner space and beyond reach of current science—mainly because of lack of imagination and intuition, arrogance, and educational brain-washing.

We can also see from this analogy how dating of the origin of the universe can be underestimated. If we imagine that the time interval between the growth of the trunk and the final twig at the end of the branch or fractal system is considerable we immediately realise that the origin of the twig is not nearly the same date as the trunk, which is older. Thus measurements don't have to be incorrect, just the assumption underlying them.

World leading physicist Stephen Hawking has stated that even in physics it is difficult to talk about the origin of the universe without referring to the concept of god. To retain a modicum of sanity, science must postulate a creative beginning even if unknown. However, the brief analysis we gave in the section on continuous life, shows that beyond all that is quantifiable—particles, waves, space and time—there can be considered to be pure (experiential) sentience. This is a perfectly logical premise when studied thoroughly as to how it relates to observable existence, in particular, when combined with the fractal gradient described above.

Scientists have created a resistance against any kind of creative source to explain the origin of the universe, producing two major objections: 1) the unacceptable dualism, that is, the incompatibility of some magical, supernatural source against the logical quantifiable aspects of modern science, and 2) the influence of the over-simplistic religious interpretation, causing science to react further in the opposite direction.

The New Science reduces these objections to a minimum, leaving scientists only to postulate that initial sentient experiential aspect of existence, in a nascent state, of no space or time (again compatible with quantum physics, the quantum reality, non-locality).

The New Science features are in fact very favourable to scientific thinking—once scientists have thought through the logical aspects it merely remains for one to become used to the idea of a source of this nature (meaning that biased mental structures are being erased). Even one of the greatest mathematicians, Von Neumann, when approached by another mathematician, who said he didn't understand a particular mathematical principle, once remarked, 'Don't worry about it, you just have to get used to it!' We can't overemphasise the massive omission in science of a creation theory. Science creates theories on fundamental issues of existence, such as the origin of the universe and the evolution of the species, understandably long before there is established evidence or proof but avoids, in spite of the overwhelming evidence, creation theories, with the retort that there are no substantial bases for creation theories with primary causes. There is nothing wrong or unscientific in postulating a beginning on the creation side that assumes some kind of initial intelligent source, particularly as such a notion tends to blatantly prove itself, even to the point of some leading scientists commenting that it is as though there was some supreme being who began creation.

The application of such a postulate then provides a grounding for further avenues of research bringing new discoveries even if the postulate was proved in error. This is quite scientific. One then simply continues to test the theory, just as one does with acceptable theories, such as the Big Bang or Darwinism.

This is a gross omission; so much so that it should make one suspicious. This accusation is further reinforced by the fact that although probably more evidence has arisen against the establishment of the Big Bang and Darwinian theories of the origins, these theories, which were once taught as theories are subtly being impressed into the minds of media consumers now as truths.

This omission actually tells one there is an underlying programme, thoroughly embedded in the academic community, which steers the mind away from this particular solution. Part of this brainwashing is to cause people to focus on the dichotomy of lifeless sciences or religious creation—this or that—which is a mental trap. There are other alternatives to religious creation, which is precisely what we are dealing with in this book.

23.

THE ORIGIN OF THE SPECIES THEORY OF EVOLUTION

A recognition of what the 'junk' DNA means, immediately eliminates Darwin's evolution as a plausible theory.

It might be opportune here to point out that whenever we encounter a viewpoint that contradicts the theory of evolution to not fall into the trap of being pulled into dichotomies (which are very much programmed into our thinking), such as: If you don't believe this, you must believe that. In this case 'that' would be the scientifically inadequate religious view (of bringing in an all-encompassing source of creation that is a man-made god-like external ruler). In other words, if you don't believe in Darwin's theory you must believe in the biblical creation. This is also a massive fallacy in most religions; the spiritual leader of 2000 years ago known today as Jesus Christ, did not teach this. Both science and religion need to study and understand the term 'within' (which will completely reconcile the two subjects, once and for all—as described in the fractal-tree section).

So what are we talking about here? Certainly a very different viewpoint. How would the materialist and atheist academic react to the following computer analogy below of evolution? Or to know that evolution is much more the domain of the physicist than the biologist (except that biologists become physicists as their subject advances)? Or to know that true evolution is compatible with the tenets of the most advanced science on the planet: quantum physics? How many scientists are familiar with quantum physics? A minuscule percentage, and a great failing of over-specialisation on the part of our society and education.

The simple computer analogy is that whenever we purchase that new computer it has acquired 'growth' in two main features 1) expansion of information capacity, and 2) speed, or rate of information (frequency). With a little more sophistication added, this is also the primary mode of true evolution, called 'ascension' of the species, or more precisely of consciousness. We shall come back to this.

This should interest the genuine scientist as it involves a much more advanced and all-embracing paradigm for science, which can reconcile all conflicts and mysteries: religion and science, theory of evolution, Big Bang, paranormal, New Age, spirituality, karma and so on. We can in fact reduce so much more to science, far more than even most, say, biologists, or in fact evolutionists, recognise. We can bring back the mind, the soul, and the spirit, handle religion—to its followers' satisfaction if they are not embroiled in ego-authority and prejudice generally.

The discomfort and disturbance to the ego of abandoning an established theory for another unfamiliar one can be quite painful: endless repetition embeds utter conviction and can override all rational thinking. Some anti-evolutionist scientists can be shown to have been more empirical in their research than the evolutionists. How many scientists are prepared to look at greater truths when it involves relinquishing fixed and familiar modes of thought, or solutions?

We might firstly note that the Big Bang theory of the origin of the universe has features in common with Darwin's theory of evolution, such as how it began and how does it evolve from disorder, a random condition, or simply from nothing. Furthermore, mainstream science states that the universe proceeds towards greater disorder (increase in entropy); it depletes its potential, whereas Darwin's evolution requires a process from simple forms towards more complex, and therefore ordered ones—an obvious contradiction. However, we referred to the much bigger picture of reality in the Big Bang section; clearly this alters the whole view of evolution of the species as well.

As implied earlier, it is not so much that science makes errors within the context of science—measurements are generally correct—it is the relative scale of the context chosen, compared with the immensely larger creation of reality. Remember, anything will self-prove relative to its own context. This is the key in fact to science's failures, which are limitations of context. Whenever something can't be explained satisfactory the solution is to expand the context—something the ego resists fervently, that is, to abandon or change the established, secure and comfortable thought structures and face new confusions. As with the Big Bang theory, the public should be outraged at being increasingly brainwashed with Darwinian evolution as though a proven fact through education and television. Unsubstantiated conclusions are now being presented as though resulting from empirical discovery.

The continued affirmation of these falsehoods through the television embeds an overpowering conviction of truth; that they must have been proven. Why does the media support these falsehoods? If you wished to bring down a civilisation you would teach these theoretical deviations as truths. A proper study of the nature of energies and frequency patterns can prove this —remember physics is primary to all scientific disciplines.

As referenced in the section on the limitations of mainstream scientific methodology a comment of great importance was made by Darwin when he stated: 'How can man judge nature (understand) if man is part of nature?' Experiments in quantum physics drew the same conclusion regarding the observer being part of the experimental set up. All the reader needs to know here is that the mind has the ability to extend any viewpoint into a larger context than can merely scientific instruments and the physical senses (the basis of the scientific observation).

Another key comment and reservation regarding his own theory, Darwin stated that if it was demonstrated or evaluated (which it has been) that 'complex organs, such as the eye could not possibly have been formed by numerous, successive, slight modifications, my theory would absolutely break down.' In a complex system of this kind its parts are interdependent in the sense that if one part fails, the total organ fails. Current theories of classical physics and Darwinism are based on 'bottom-up' theories: everything began with the smallest parts and developed into the complex. Owing to the great advances in molecular biology and biochemistry there are thousands of irreducible complex forms, meaning they cannot be explained by giving causation to the parts. In complex organs, or even a whole human being there must have been a plan (a template of the wholeness to guide the particles or cells into place).

Of course the holographic fractal model already outlined, totally demolishes the classical physics aspects and Darwinism, but so does quantum physics. The parts-to-whole mechanism (bottom-up) would be expected to automatically incorporate compromising principles, such as 'survival of the fittest,' to remotely suggest a viable survival requirement of a type of evolution. Nature has simply done its best (resorting to compromised strategies) with a grossly impaired evolution from a hugely mutated DNA of all life on the planet (the 'junk' DNA).

The mechanism is that once a functional advantage of a member of a species developed, its offspring would inherit that advantage and pass it on. This would be a very gradual and slow process as Darwin recognised, and the less able members of the species would die out and the superior members survive. It is competition. A lot could be said about competition, the psychology of which hasn't been studied on this planet—in its recent evolution. It is not what it is made out to be, which can be proved through the physics of energies or frequency patterns that true intelligence and evolution is towards greater integration—not exaggerated separation and subsequent fragmentation. The point being made here is that there are far better systems of evolution, even though we are not exactly demonstrating them.

To continue with natural selection, as implied, this is the bottom-up principle or parts to wholes—the part level is the cause and the whole is merely an effect (rather than the other way round). We may agree that the natural selection rule and subsequent role of functional advantage could explain dominant persistence of some simple forms over others, but each form becomes more complex, exceedingly complex in some cases, and the functional advantage program of a species evolving a complex organism, functioning as a whole, can't be explained by trial and error selection—to say the least the improbabilities are far too great to arrive at such high organisation. One simply cannot develop a higher order from a lower order without at least initial guide lines.

Nevertheless three established gaps in the theory are 1) how to explain spontaneous life from organic material, 2) the gap between animal and vegetable life, and 3) the gap between different species. Moreover, there is no evolutionary sequence amongst the diverse cells on Earth. Regarding these gaps, if we replace material ('bottom-up') motivation with consciousness ('top-down') these three problems are handled.

Molecular biologist Michael Denton, in his exceptionally well-researched and well-written book *Evolution: New Developments in Science are Challenging Darwinism*, with no religious bias whatsoever and who in fact considers Darwinism the only true theory of evolution (that we have at present), elaborates with substantial evidence the untenability of the principal features of Darwin's theory, such as the features of gradual transformation, and the development of complex forms from simple forms. Darwin's evolution holds that all life is related and has descended from a common ancestor. Oddly enough the New Science could completely agree with this, even the word 'descended' rather than 'ascended'. However, the evolutionary medium would have to be 'consciousness' itself—which is made up of a complexity of frequency patterns, much more elaborate than physical forms.* The above statement would thus be in accord with the New Science if it were saying everything (life and the universe) all descended (top-down) from a single source beyond space and time (the apex of the fractal tree hierarchy). Thus real evolution is ascension (following creation)—a process of consciousness integration through the many layers of the multidimensional structure (one might envisage the twig 'ascending' back to the trunk). See Figure 10.

What about the popular comment in support of Darwinism that the number of genes of the ape is similar to those of humans? Firstly, the more recent count of human genes, which has reduced considerably over the years, is around 25,000. Secondly, the fruit fly has about 15,000 genes, and even more amazing the minute round worm with 1000 cells or so (compare human at over 50 trillion cells) has about 24,000 genes. Thirdly, regarding the ape and human genes comparison, only around 300 of these 25,000 or so are relevant to this comparison between the ape and human, and 223 of these manifest the differences—which is huge relative to about 300, and thus demonstrates very large differences. Moreover, according to archaeological artefacts and historical records some animals have remained unchanged for over 10,000 years to this day; for example, domestic animals. [* When expressing through 'mind'.]

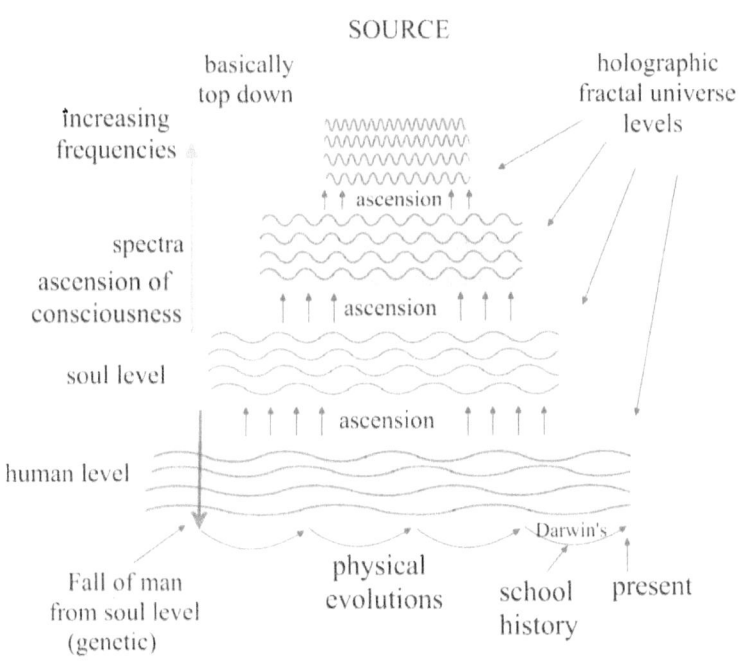

FIGURE 10

Darwin's letters indicate that he was a relatively modest man for an academic. He was not over-dogmatic about his theories and was ready to accept the observed discrepancies. It was the scientists and authorities, whose need for a simple answer was fulfilled by Darwin's theory, who clung to this inadequate theory. If Darwin had been alive today there seems little doubt he would have abandoned his theory, in particular, when one recalls his comments—some already given—for instance, '... to suppose that the eye ... could have been formed by natural selection seems, I freely confess, absurd in the highest degree.'

This is reminiscent of Einstein's relativity, which he considered incomplete or in fact wrong. It was the relativists who pounced on this clever theory to reinforce their authority. (Unknown to most, however, there is a type of covert politics manipulation (of education and minds) to dumb down the population with the most de-graded solutions to life's most important issues—one example of an exposure of this is available on a website that has a list of professors' names all protesting the Big Bang theory, stating that it only came into acceptance due to government funding.)

Regarding the problem with complexity being inadequately explained from the particle level or bottom-up process, how do even more unitary experiences occur, such as different areas of the brain linking together? Quantum physics could handle this with nonlocality even in examples of stimulus-response, such as the classic one of Pavlov's dogs, regarding the association of the stimulus of the bell sound and the response of saliva glands. Orthodox science posits that the stimulus and the response can be connected mechanistically, that is, associated in the brain. However, as per the New Science there will be a whole quantum state underlying the two separate functions, which is a mind ability and experiential. The mind oversees the separate parts and brings them into unity. Investigators might consider what happens when the conditioned animal is presented or merely tempted with food; will it hear the bell in its mind? It should. In the later more technical section we shall see that the quantum state (in this example) is a wave function carrying the waves or oscillations of both the stimulus and response (remember, each wave is an undivided whole).

All life on this planet is in an extremely non-harmonic and parasitic evolution. Thus there is some truth in the evolutionists' findings, since our history reveals that much in life and the mechanics of the universe are malfunctioning, leading science astray in evaluating these flaws as truths—recall the far-fetched examples of our Milky Way black-hole, in a degenerative condition and the erasure of all cellular memory evidenced by the 'junk' DNA.

Though not generally realised, evolution is a field of research for the physicist, as alluded to earlier—it is the subject of integration of intelligent frequency patterns, involving an ascension process through a multidimensional, holographic, fractal structure into greater realms of expansion (wholeness, 'holy') and higher rates of information (frequencies), eventually (after maybe millions of years) going beyond all creation back to the Source. Everything ultimately returns to Source, and another colossal plan will presumably be set forth, and in the process the One Self, Source, is forever becoming.

It is logically impossible for creation to occur either from nothing or, once begun, from the local (particle, bottom-up principle) to the global design of order and interrelationship between the parts. The mind constantly falls into the trap of unconsciously making assumptions. How can intelligence develop from no intelligence without a guiding principle; it is impossible. The precise flaw in the thinking here is that the mind is referencing its own assumptions (it is biased) either by subconsciously believing they are truths, or by taking the assumption as a zero point of reference (when it fact it has a value and is a context that influences the outcome). Materialistic and atheistic scientists ironically don't have to fear the concept of an all-intelligent source, which would only need to provide the basic (higher-dimensional) blueprints—these are templates for shaping energy in space (and time) that are allowed to evolve, and all the energy is of the single Source. There is enough science to discover to keep scientists going for millions of years—recap on the three examples of advanced knowledge given earlier.

The big puzzle amongst the many scientists, even who specialise in the relevant fields and who have arrived at a definite conclusion against Darwinism, is how could a theory, lacking insufficient evidence become such a convincing dogma. Clearly it was a hugely needed validation and relief for many scientists, securing their authority even over theology, once and for all it seemed. Dogma is generally accompanied by intellectual insecurity, arrogance and ego, generating a compulsive need for an all-embracing theory that one understands, but which will always have assumptions that lead to belief systems. These embedded programmes become utterly incontrovertible facts to the individual harbouring them, once the subconscious dominates the conscious. This is not a puzzle in the New Science, but further explanations are beyond the scope of this brief book. Recall that 'moulding the mind' is literal—the universal computer system functions on geometrical intelligence.

The main objection from science, regarding evolution theories (and from scientists not biased by religion), is firstly the lack of scientific verification for a transitional continuum linking together all species, and originating from individual cells, and secondly our usual criticism of order and design appearing from a random condition. Typology or the principle that different species are isolated, unique types (modifications due to interbreeding is a limited argument again this), has much more support today than an evolution by gradual development from one species to another as required by evolution theory.

There are also variations on Darwin's theory, such as neo-Darwinism, which takes into account progress in genetics, and Quantum Evolution, which was proposed to explain the missing links and missing evidence in the theory of gradual evolution from one species to another. However, it makes little difference to the drastic change in paradigm that is required to handle the major issues such as the beginning and where the order came from.

A valid and workable theory of evolution must be based on an understanding of the relationship between the part and the whole—which would resolve most advanced confusions. Classical science—physics, biology, chemistry, etc.—doesn't even recognise there is a relationship—quantum physics qualifies as the closest.

24.

THE MIND IS A BY-PRODUCT OF THE BRAIN

Can a brain research and develop a quantum computer?

Mainstream scientists have been taught, and therefore most believe, that the mind is only a by-product of the brain, which means of course that at physical death the mind is extinguished as is the brain. Science will continue to find lower mechanistic explanations and attribute them to brain activity. Most scientists have been educated to believe that consciousness is an emergent aspect of neural networks. Nevertheless as they go deeper into the subject and more remarkable mechanisms are discovered about life and mind, the question keeps popping up as to whether a by-product of something, such as the mind being considered as a by-product of the brain, can be superior to the product. To even entertain this is logically unsound. The only way it can work is for the by-product to already exist (as energy, but not non-physical software). However, at present, science can't detect these energies, just as it can't detect the source of virtual particles (and doesn't even recognise that there is a source).

The brain is a machine: it processes symbols like our computers. There must be something sentient, something capable of experience, underlying the gross mechanics. There is no dilemma here, as scientists believe, of the apparent requirement of machines controlling machines and other machines controlling those machines, and so on, repetitively to infinity. There is a logical progression integrating towards the non-space-time condition of wholeness; that which experiences, that is, pure consciousness.

Mechanisms in which the whole equals the sum of the parts can never create true unity or the quantum state. Something must span the parts in order to link them—even in stimulus-response mechanisms. All natural entities, atoms, planets, stars, universes are undivided wholes (at some point within the inner space of the holographic, fractal configuration). These are single quantum states (with oscillations not remotely detectable with current scientific instruments of low resolution). If scientists studied the arts as well as the sciences they would begin to grasp the nature and reality of unity (art is all about unity—true organisation—which society is losing sight of).

The major part of existence is experiential; clearly this would be avoiding the issue if we regarded this as an illusion (non-physical or a by-product). The materialistic approach (Darwinian and Big Bang) is to consider everything as built up from particles with no guidance as to what form these particles are going to take. Any intelligent program that exists in this development must have been put there.

Materialistically-driven configurations of this kind are simply particles stuck together by forces—they are part of the structure of matter, space and time. The concept of 'idea' cannot emerge from this, since it has wholeness (it can't magically pop out of the virtual state or quantum reality as though from nothing, since these are energy levels already there. This (idea) is a quantum state of energy and inherently does not contain parts. In the process of cognition, connecting pieces of information cannot attain integration of wholeness without making a selection from the *unmanifest* potentia, or a higher fractal level already manifesting the idea.

From the section on continuous life, however, we identified the fundamental medium, 'sentience,' or the primary state of consciousness, which has no location, no space or time, particles or wavelength. Consider the following simple example: two people are communicating using their language of words. As one receives each word of, say, a sentence from the other, the meaning or understanding does not build up gradually. When the words have been transmitted, or sometimes just before completion, the recipient instantaneously receives the cognition.

The Emerging New Science

Underlying each word there is a unit of energy, a quantum state, and similarly for the whole sentence, though it may contain more than one single group of words conveying a single idea.

It is of interest to mention that the creation of quantum states has been investigated for about 50 years. Scientist professor Terry Clark designed the first device to produce a quantum state of about half a centimetre in diameter in a small electron ring. This means that when the ring circuit received a signal, the signal instantaneously created its effect everywhere in the circuit. Since then, quantum states a metre in diameter have been investigated. More recently this research has been extended to the application of producing faster computers. Furthermore, computers are being investigated today to utilise quantum mechanics principles, such as superposition of quantum states. Science is not recognising the extent to which the mind functions on quantum principles. The mind can operate fully nonlinearly and holographically, which means there is extensive use of superposition and also super-imposition. The brain cannot achieve this form of simultaneity, superposition and 4D holographic programming (see later sections on collapse of the wave function and quantum computing).

Science and education, and usually the media, are making an excellent job of programming the population to believe that man is just a brain and a body. Even though the mind did get some early acceptance as an entity in its own right, particularly in philosophy and religion, science has accommodated this so that any mind phenomenon, such as an energy-field system is merely a by-product of the brain. Unless some drastic change is made in the educational system, this will be the majority belief system within the next generation; that the brain is primary and mind, secondary, or an effect of the brain. This is an example of how degraded our knowledge has become on this planet. Most people today, however, including an increasing number of scientists believe that mind and consciousness are valid structures apart from the brain and working with the brain; the latter is an essential frequency step-down device between mind and body.

Let us introduce an observation that anyone can make, and in fact that has already been made unknowingly, which should give the reader proof that mind activity is beyond the brain and is superior. Remember that observations are generally all that is required for 'proof'. There needs to be sufficient agreement though regarding what one sees or thinks, and this unfortunately is governed by the individual's perceptions.

We shall give a couple of examples to illustrate that the observational evidence is present at all times, but only the second example will be sufficiently amenable to evaluation to make sense. Under these imaginary conditions of a body operated by a brain only, if we try speeding up our walk or even jogging over uneven ground, we would fall over. The feet at the ankle joints would not be able to adjust rapidly enough to retain balance. There certainly would not be any tight-rope performances. Unfortunately common sense is not sufficient to evaluate this claim.

The second example should be straight forward, as it blatantly reveals this evidence of mind activity as opposed to solely brain; it is in keyboard playing. If we are only a brain and a body, then it takes time for signals to transmit from the brain to, say, the finger tip. We can take this to be around one-tenth of a second. This is a long time. Thus to control a finger muscle there is a delay of about one-tenth of a second before the muscle can be caused to act, or one-tenth of a second before the brain knows the finger has moved (by means of the feedback).

It doesn't take much imagination to see that if a concert pianist is playing at high speed, say, 20 notes per second, with a delay of one-tenth second, he or she will not know which finger is moving. Motion would constantly stop for relocation to take place—a little like going down stairs hastily and misreading which step one's foot is on; consciousness must re-anchor itself to the precise movement and position in relation to the environment. But we can know that a concert pianist is in control of each finger to a high degree of precision. Therefore control is virtually instantaneous.

Thus, astonishingly, in terms of current theories, the communication system to the muscles (or joints) is virtually instantaneous in learned movements. The mind works on quantum action—instantaneous states—and thus can easily handle this problem of control over movements at high speeds. (Clearly a field system is in operation here.)

Another problem with the claims that the brain system is in charge is to do with information capacity. The artificial intelligence expert will know that the physiological information capacity is much too low to provide skilful movements. What we are pointing out is that the number of motor units, or independently operating muscle-fibre groups, is totally inadequate. One might figure out a satisfactory design of information storage in the brain, using all the permutations and combinations one can muster up, but even if this information density (of 'bits') was adequate, it has to pass through to the muscles which act like a bottleneck.

The huge limitations on what the muscles can receive will filter out the information density from the brain, corresponding to an inadequate physiological level. There are only about 250 million muscle fibres in the whole body and although each fibre contains thousands of fibrils and filaments, they do not according to established science act independently—which is what we are interested in. In fact, only groups of fibres of varying number act independently, that is, as a whole—as one. This is like, say, giving someone the task of drawing on a sheet of paper an outline of, say, an animal, and they are given a number of pins to stick in the paper to form the outline. Let's say they are given a million pins. This corresponds to the amount of information (bits as in a computer, available for this drawing path—the latter is an analogy for the skilled-movement path). However, a template is fixed over the paper, which contains a relatively low number of holes in it for the pins. In fact, let's say only about ten holes coincide with the predicted drawing outline of the animal. One can now see that the result of ten pins outlining the animal would render it unrecognisable, or we can say the drawing path is very inaccurate.

This is resolved with a theory of mind in which a field system (containing the learning patterns and programmes) exists around the joints and linked to the muscles, regulating the angle of the joints. Thus we now have a system which can act instantaneously and provide any degree of information density necessary—even for the highest imaginable skills; information storage is potentially unlimited.

Another problem with existing knowledge is the belief that the information capacity cannot be increased since there is ultimately a fixed number of brain cells. This will create a ridiculous limit on potential abilities. It is like having a computer and one cannot add extra memory chips (or get a bigger computer). There is no problem here with the mind, which is an energy structure. Just as cells divide in reproduction the electromagnetic units of the mind, which are much smaller, multiply by fission. Two units 'grow' from one, but the one remains (a natural characteristic of basic creation and consciousness—see later sections on creation). It remains on a slightly higher-dimensional level or frequency band and is intrinsically connected to the two it has born. It keeps the two in a proper relationship to one another.

This is not so different from the idea of a supervisor hiring two workers who now do his job, but he monitors them. The learning pattern can be expanded in this way, increasing the information capacity (like adding extra memory to the artificial computer). This can be achieved by special exercises designed to encourage the mind's ability to adapt, somewhat akin to the process of muscular development in which the individual, employing resistance exercises, tears down the tissues in the muscle—in effect, creating a discrepancy, a gap for the body's resources to repair this and bring further development. Similarly the exercises for fission of the electromagnetic units creates a 'gap' from the discrepancy of deliberately reducing tension in muscles, coaxingly and repetitively, to values just below the threshold for the full desired movement. That is, introducing intentionally too much relaxation but at the same time desiring the complete unimpaired movement. The mind handles this by increasing the units—the 'bits' in the computer system, which then require less effort/energy to produce the same result.

Another important area of muscular activity, which is associated incorrectly with brain processes rather than mind, is the resistance that arises in physical movement. Science recognises some mysterious resistance in addition to the usual classical resistances of inertia and friction. Science will find that this is not physiological resistance at all, but is due to the mind computer system regarding limited rates of information from the learning patterns. These frequencies can be increased by special exercises. This is not a property of the brain but the mind computer system, which can be expanded indefinitely along with increasing rates of information, such as for higher skills.[5]

In simple terms, then, we are stating that if a muscle is in good shape but communication to it is inefficient, the muscle will not operate as well as it could, or if the efficiency of communication to the muscle is very high it will make full use of the good muscle condition. And vice versa, if the muscle is in poor shape but the communication highly efficient, then the communication will be wasted on it.[6]

Universal evidence demands that all natural entities have true unity: 'unbroken wholeness' as Professor Bohm calls it or, in other words, are whole quantum states. The brain can't produce this. Returning to the application of quantum states to instantaneous perception of movements we shall find that so-called signals from the brain to the finger, exists within vortices (spheres of energy) and programmes around the joints, so they don't really have far to go, even if delay times were involved, which they are not. These quantum fields are part of the mind system and function instantaneously.

Notes
1. Michael Roll, www.rense.com/general32/sub.htm.

2. www.nhbeyondduality.org.uk. Article, *There is No Death: You Can't Die if You Try.*

3. www.nhbeyondduality.org.uk. Article: *The Two Hidden Stumbling Blocks Inherent Within Current Science.*

4. Ibid. Article: *The Philadelphia Experiment.* Also book: *Engaging The Extraterrestrials* by N. Huntley.

5. www.nhbeyondduality.org.uk. Refer articles under 'Skills'.

6. Book, *The Attainment of Superior Physical Abilities* by N. Huntley.

PART FOUR

THE NATURE OF CREATION

25.

THE UNIVERSE, LIFE AND EVOLUTION, AND THEIR PURPOSE

Manifestation and exploration of the infinite possibilities

In the early sections we covered some of the extended features of the universe, in particular, revealing the enormous difference between what is taught on the subject of evolution, and the true meaning, better described by 'ascension.' Here we shall clarify some of these points and elaborate more on the purpose of existence.

In the section on continuous life we saw the results of the thought experiment to achieve the nothing state by gradually eliminating all structure, all that is quantitative and quantifiable. We arrive at (or it was necessary to postulate) the state of sentience. This is the native-state awareness, the quintessential essence of 'aliveness' and ability to experience. It is beyond space and time, there is no energy, particle or wave, and it is not quantifiable and never will be. It has no location and is eternal. It is unmanifest potentia. It naturally has true unity, and can only be an undivided whole containing potentially infinite possibilities. It can be considered to be totally subjective and infinitely nonlinear.

We have mentioned the ladder analogy for the different levels of the fractal universe or multiverse (each rung represents a universe layer). We can also use the tree analogy (fractal tree section), which is a more advanced and accurate model of creation than the ladder.

Nevertheless as we go up the rungs in the ladder analogy there is an increase in frequencies and dimensions. For the layman, just think of these dimensions as wavebands of increasing frequencies (and the waveband as a row of wavy lines). Also integration of energies is increasing as we go up the ladder. This means harmony, wholeness, less emphasis on polarity separation; in terms of behaviour, less ego. There is more perception of the whole. Since everything is made up of consciousness (sentience) and its fundamental particles, then everything can be potentially known by a state of beingness; that is, by resonating one's consciousness (and duplicating) to achieve direct understanding. Remember that when two frequency patterns match (such as consciousness duplicating an object) there is quantum regeneration, a single state is achieved and there is a sharing of the information being spanned. This is the ultimate way of understanding, in effect, by direct perception, of internal consciousness upon external consciousness (meaning by 'external', consciousness within all material things), and taking on the original thought of creation. However, this ability is so atrophied in the human race it is not even generally known as a possibility.

The purpose of life and existence is primarily the process of the Source (the original whole) knowing itself by expressing 'outwards' and objectifying different levels of existence. This involves different degrees of separateness (increasing as we go down the ladder) for the exploration of consciousness by experiencing its manifestations through degrees of individuation. Lower levels can evolve or ascend by waking up to the fact that everything originated from the same creation. The tree is a good analogy for this. A twig may appear separate on the first fractal level of twigs but it is clear it connects up to everything else as we go up the branch. Ultimately everything connects and goes back literally to Source, which through this process is forever qualifying itself and becoming. We could say that true evolution or ascension is about creating qualitative states (quality) from the quantitative—putting order into the parts, bringing about integration (through quantum regeneration), expanding consciousness and return ultimately to Source (after millions or billions of years depending on the consciousness).

Today, this ascension principle—of Source projecting out expressions of itself, endowing them with appropriate limitations but programmed with a freedom to evolve back to Source—is complicated by life forms returning in reincarnation to fix their mistakes (not part of the basic blueprints but always inherent through free will and infinite possibilities), which would cause them to descend; hence karma—see later sections. Note that ascended beings may incarnate on missions to help a civilisation.

    The reader may have noticed what might appear to be an inadequacy in the use of the analogies if they are used for illustrating ascension. For example, using the fractal tree, we have to imagine the twigs as merging into the branches, which merge into larger and fewer branches, ultimately becoming one in the tree trunk. Or better still in this case, take the analogy of the company organisation. How can all the ground-floor workers attain gradual promotion (ascend/evolve) to become the president—one president? This is logically impossible. But the logic of the old science is merely 3D, external and linear. It cannot handle infinite (internal) non-linearity. This is also one of those instances where people are educated with viewpoints about as opposite as they could be to truth. We shall give the New Science version of this and at the same time increase the reader's understanding of the mysterious primary state of sentience or pure consciousness. This will seem initially totally unacceptable to the ego, and we are taking a risk presenting it. (In the author's first book *Superhuman*, over 30 years ago, this information was deliberately omitted in the last chapter on the God concept, since it was considered it would never be believed.)

    Thus we need to explain merging of consciousness as a natural process of evolution. The following example is for illustration purposes only, though theoretically it might conceivably be possible between two twins.

    Consider person A and person B. Let's assume they bring about a high level of attunement between their respective consciousness. This could be due to long associations, past life connections, same origin, twins, or direct manipulation of genetics for correspondences. Person A and B are now thinking so much alike (partially or totally) that their similar consciousness can merge into one whole (in quantum physics, this would be a new whole, a new quantum state and wave function).

During this merging, each person doesn't observe the other's mind as being separate. We are not talking about a few signal correspondences or synchronicities but the matching of a large part of the individual's frequency spectrum and coding. Let us take this to the extreme and suppose in the case of a very high degree of attunement, B volunteers to move from his body into A's mind and body. The two minds form a new mind C; not A and B attached to one another. This means A feels no intrusion; in fact feels just like himself or herself with various enhancements additions, memories, abilities. But what about B, where is B?

Person B feels exactly the same; in a different body though. There is no intrusion. It feels just like B; to person B. The simple reason for this is that everything is formatted (is a program) from the unified field—of infinite units (particles) of consciousness. It is the same consciousness.

If we wish to retain some difference, such as between A and B, that is, not all thoughts correspond, then each will retain this personal area unknown generally to the other, though more accessible to each if in the same body, since as soon as these different thoughts affect the common areas, such as moving the body, then the other would be in immediate attunement with these concepts. In fact a rapid adjustment would probably take place so that all extraneous differences that might conflict would harmonise and consolidate into common fixed patterns of behaviour.

In the positive sense this type of merging is what occurs as we ascend (the 'twigs' merge into the branches, which merge into 'higher' branches and eventually all merge into the 'source', the tree trunk). In reality the Source would then 'project' out (breath out) another great plan for further exploration of its consciousness. However, in spite of this organised structure for ascension there can be variations. What if one's progress is held up by, for example, one's parallel twin (which may be in a descending mode), or other laggard members in one's group? According to the Guardian material, if necessary it can be sufficient for ascension to simply acquire a copy of their codes (encryptions)—the appropriate quantum of energy will be drawn into it, enabling a more 'sovereign' individual to ascend.

The Emerging New Science

But how far this method can continue is not clear, without the regular merging.

Following on from the section on continuous life, searching for the 'nothingness' state, we arrived at pure consciousness, unformatted sentience, which is the experiential ability manifesting in all life activities. 'Sentience' then must be able to objectify this primordial experiential state. Amazingly the physics of this seemingly paranormal, supernatural requirement can be modelled by quantum physics—the quantum reality has some corresponding characteristics—in particular, there is no true or absolute objectivity, only relative. Part of the primary Source separates from itself. The Source itself as One can only be subjective; separation is required for objectivity. The subjective state of infinite potentia and possibilities is like the quantum reality (of phase-correlation) that quantum reduces, say, a particle (phase-randomisation, such as what we observe). Many particles can then be organised and formatted, held in a pattern (mould) by a larger quantum wave by means of the creation of a template (blueprint).

There can only be harmony in the pure sentient state—anything less would cause one part or parts to divide from the rest. We can see that the term 'infinite love' (of a Creator) is quite logical—all potential parts in perfect harmony.

The above is a perfectly logical theoretical approach. We postulate that the sentient state focuses into a tiny part (particle) and must, of course, separate from that part to manifest an objective form. We now have the first signs of space (if there's a location, there must be space) and time (when there is change).

To do this it must create separation within itself, so that one part can observe another—even to let one part not know another part. The introduction of limitations in this way creates the separateness characteristic and quantity itself; and subsequently the quantifiable and objective condition necessary for self-observation.

Note that—and this is very important today—even the objective form, such as the particle, is not actually outside the source. It can only be inside (there is nothing outside Source even today); it wouldn't be rational to think this on the basis of what we have just explained, hence the vital word 'within' repeatedly used in the Christian Bible but not generally understood. The fractal tree gives a very simple analogy—the twig (the part) is quite separate at its own level from the trunk yet it is connected internally further up the hierarchy of branches towards the trunk.

One might object and point out that the twig is outside the trunk as source. So why is everything inside Source? However, recall that this tree analogy is in 3D and if we looked into the inner space (within) of the tree, we would see spheres within spheres (higher-dimensional vortices). The trunk would be the largest sphere, with branch spheres and twig spheres inside it of decreasing size (before it projects out externally into 3D).

This particle is a speck of sentience—but structured to hold together as a particle. Imagine a drop of water that must be encapsulated by a fine membrane to hold its form otherwise it would (let's say) immediately vaporise into the surrounding vapour (an all-encompassing unified field). But remember, the encapsulation itself is made of 'condensed', 'solidified' sentience —what we call actual energy.

Let us recap on what we have covered and pursue this further. Ultimately everything is built from this same substance, sentience, which is also called consciousness. But it is pure consciousness, that is, not formatted, not of, or dependent on space and time, has no locality, is eternal. It is a quantum state; an undivided whole unit. Nevertheless the particle of sentience has been separated and thus must have a boundary; thus it is now formatted or moulded consciousness to retain a particle or wave-form pattern.

Therefore, in a sense this separated consciousness is more limited. It is formatted, and has been made finite (has been 'shaped', since creation works on geometric intelligence). More particles are created and held into groups by larger quantum states (particle/entity). See Figure 1. These states must be in continuous creation. Even if put on the automatic, this mechanism itself has to be created. But since there is no such thing as continuous—this involves time—these quantum energy states 'flicker' on and off. They can only change by means of a new frame. In order to create time, something must switch off, then on again, say, to produce duration or change, and subsequently time. And we have frequencies. The quantum states, such as particles, are created and uncreated repeatedly; somewhat analogous to our digital computers with their on/off electrical pulses. [Strictly, there must be a background quantum state; more on this later.]

Patterns of frequencies are set up that can be considered to have geometric encoding (geometric intelligence). The basic thought process, from the eternal sentience, projects out these patterns as blueprints or templates (like a mould), condensing a rigid framework, that is, from unmanifest to manifest. [Our thought processes apparently form templates which then accrue matching frequencies (of thought), building up a stronger manifestation or thought form.]

In separating from itself, the Source can now make a selection from the sentient state of infinite possibilities within itself, the highest fractal level will be created first, higher dimensionally, and containing many probabilities for the lower levels to select from. Countless particles, now having been created, are projected through these templates (structured of the same particles) to give shape to various forms. Just as this first particle is a unit of consciousness, all the larger wholes (underlying atoms, cells, planets, stars) are larger states of undivided consciousness—undivided in its larger state (Figure 1(b)).

As a slight digression let us add that recognition of these characteristics of consciousness as opposed to those of machinery can be corroborated by training the mind in observation of itself. We mentioned the example of verbal communication, in which understanding is apprehended in whole quantum states—cognition, not parts/elements built up. A further example would be in the visual field. There is a wide field of view that we are conscious of, and although it appears that the intensity of attention (consciousness) is greatest at the focus of the eyes and fades towards the periphery, in fact, there is a rapid superimposition of quantum states ranging from the whole field of view to the point focus flickering on and off, giving the illusion of a continuous gradient of intensity. There are countless examples. Within languages, letters, words, phrases or even sentences are whole quantum states (waveforms).

Note that we are now using the word consciousness. It now contains particles (much smaller than subatomic particles) and wave patterns (that is, particles in unison forming waves that interact with one another). The state of sentience itself is nascent, a native-state awareness with no modulations (wavelength, shape, moulding). However, in the expressions of shapes, which also means limitations (giving rise to space and time), we can call it consciousness—a specific directional formatting of the sentient experiential mode. That is, we are still calling it consciousness even when it is directed, and 'shaped' to convey information, but we could now, due to the information assistance of the waves and particles, call it 'mind'. Thus pure consciousness or sentience is beyond space, time and particles and is a portion of the absolute, but in creating particles and waves, it combines with them to form mind.

A whole system of dimensional frameworks could be set-up (compare fractal tree), and because of the nature of consciousness this multiverse structure automatically expresses unity, or infinite nature, in limited and selected quantitative ways in a fractal system but of which the parts and the whole are holographic. The holograph principle is a mechanism endeavouring to replicate the infinite nature of consciousness, of which the latter, since it has complete unity, then every imaginary point within it and every sub-whole are all everywhere at once (but at the original sentient level).

Imagine a region of space, say, spherical, which has total unity within it—a quantum state—then firstly a ray of radiation entering it would immediately be instantaneously everywhere inside this sphere. Secondly, there would be no left or right, up or down, or front or back—every point is everywhere at once. The state of perfect unity would be infinitely nonlinear (see later sections). Note that quantum states have been created artificially by scientists, as mentioned—the first quantum state, half a centimetre in diameter was produced by Professor Terry Clark about forty years ago.

One might see now why the holographic expression arises. It is a mechanistic, quantitative description of the eternal sentience, which also has the property of the whole within each part. Thus fantastic mechanisms like holographic and fractal systems, which are observable, are clues and manifestations of the infinite nature of 'sentience' or consciousness. Hence whatever creates these external manifestations must be equal or greater in information. We could also say that the holographic fractal mechanism is the interface or link between the absolute and the relative.

Why is our existence fractally organised? If it wasn't, there would only be the smallest particles and the wholes—no molecules, cells, planets, stars, etc., which are whole fractal levels in themselves. Thus the properties of the holograph and fractals give us something tangible, observable, as to how nature creates.

Now the templates or blueprints would be higher-dimensional, which means they can hold many probabilities for selection by evolving life forms. However, they will have rules programmed into the fractal level, specific to certain requirements, such as to explore the objective states/dimensions created from the subjective sentience and degrees of free will. A higher level of consciousness—for example, an advanced being—will select the higher fractal levels but may ultimately explore further ('downwards') fractioning, giving rise to lower frequency spectra and lower life forms to be explored by creation.

There is a danger though, as further universes are created on lower-fractal levels, which is like the tree trunk extending its branches to still smaller ones, the perception of separation is increasing. The fractal twig level is the most separated level (fragmented). This level has smaller quantum states, there will be less perception of unity, and greater sense of objectivity, with a degree of separation from, or not even related to, the subjectivity—'the environment is out there and nothing to do with me' (a typical belief of the lower levels), ideal for experiencing effects, good or bad. However, as alluded to in the section on evolution there was a similar 'extension' (of life) to a lower fractal level (4D to 3D) but this wasn't part of the universal plan (of Source). This was a chaotic fall; the so-called (genetic) Fall of Man.[1]

What is the connection between consciousness and the actual structure of the fractal, holographic multiverse? Originally portions of the sentience are separated out by means of the necessary template and boundary conditions for isolation from the whole. These may be vast gestalts of consciousness and have been called monads, which further divide.

To recap on the simple analogy given earlier, just as white light is split into its various colours, by means of a prism, giving rise to our well-known spectrum from the highest frequency of violet to the lowest frequency red (in this short spectrum), the monad divides up its frequency spectrum, separating out into dimensionalised wavebands (like the colours) and projecting them into the fractal layers of the multiverse structure. The 3D spectrum would be in the lowest spectrum of the first fractal level; compare twig in fractal tree. This would be the lowest aspect of consciousness appropriate to the human level at this time. The next portion of the monad and group of frequencies would go into the next fractal level, which is the soul level (4D and above); and so on back to the final fractal level and maximum integration, corresponding to the tree trunk, the Source. But the procedure would divide from the top down, such as branch to smaller branches, to many twigs.

Another big question might be, How can we justify calling it infinite? We could answer this by referring to the incredible creations in the universe, but this would be putting the cart before the horse.

Let us use the same mathematical and geometrical conceptualisation as we did above as a possible application to this problem, that is, the sphere with total unity within it. Now recall that we have already proposed that life and the universe functions on geometric intelligence (has geometric encoding systems). As previously stated, as soon as an appropriate signal enters this region it is everywhere at once. A quantitative interpretation of this would be it is infinitely holographic. Every infinitesimally small and imaginary part and every subpart of every conceivable size is connected to the whole and every other part. Thus the holograph is a good model.

We already know that if we take a hologram photographic plate, which gives a 3D image when a laser is beamed

The Emerging New Science

through it, that the full 3D image appears, even if we cut the plate into smaller pieces and shine the laser through. Thus true unity could be considered to contain infinity; is infinitely nonlinear and that infinity is not across 3D space but within inner space. It is interesting to note that if we imagine within this sphere of complete unity, an infinite number of infinitesimally small points (spheres) and then similarly every other size, all superimposed, we might see that in different combinations every imaginable form could be shaped.

The reader may be familiar with (even weary of) the expression of increasing use, 'All is one,' repeated frequently, in particular, within the New-Age movement, almost as a panacea for all negativity. One may then question this by pointing out that how can this be if just about everything we face daily involves separation, division and isolation. This is where the dimensional layers come in. They can't be external though (across 3D space). They must be nested internally within inner space—the actual domain and passage of creative energy as it perpetuates the dimensionalised fractal levels.

If we use the tree analogy it becomes easier. We can't claim 'all is one' if we take it that, say, existence at the twig level is the only level, since the twigs themselves are clearly separate from one another. But taking into account the whole, we see the connectivity. Thus there must be recognition of the branch system integrating towards the trunk. We can then see that although the twigs are separate at their own level, they are all connected through the complete tree. The same applies for the branch fractal levels. If one focuses solely on one fractal-level context its separateness is valid, but it is also correct to say that it is connected if we shift the context to a higher level of the whole tree. One can also see that a consciousness spanning many dimensional fractal levels can experience separateness and unity simultaneously. Let us finalise this section with a further argument supporting—in fact demanding—that unity and interconnectedness must be a basic condition underlying all creation.

Everything exists as frequencies, that is, on and off, this gives rise to time. Imagine the first 'on' state of creation—it is a whole quantum state of energy. If it simply flickers off and is replaced by the next 'on', there is no relationship with the first 'on'—it might as well be the first 'on'. There is still no time because we didn't say anything about the first quantum state being recorded or remembered. In order to give the appearance of time, these quantum-state pulses must be on a lower (dimensional) level from their source, of which the latter must still remain. In other words, there must be a larger quantum whole that unites, spans, the two lower-quantum pulses, in order that the second pulse registers within the primary quantum state as different from and displaced from, the first pulse. This also forms the basis of the triad principle. One can see that, even if briefly, for these two fleeting pulses, a small element of time is created.

By extending this hierarchically, a fractal system of space-time is created. All apparently isolated particles, entities, or quantum states must have unifying origins, which are contexts from which they came. As mentioned earlier, even the connectivity of a stimulus and a response, has a third factor that is the unification of the stimulus and response—a single quantum state.

Thus logically we can see that there can be no existence without organisation within organisation, the fractal hierarchical, holographic configuration. Science's model of a one-level (just a 3D fractal) description of existence can't ever explain the properties of that existence that we experience continuously.

26.

DUALISM AND THE ANTHROPOMORPHIC PRINCIPLE

The universe and man came from the same source and any apparent dualism disappears.

Dualism is the apparent presence of two incompatible concepts, yet seemingly working together, such as the body and soul, the physical and the spiritual, or the bottom-up versus top-down principle. Dualism is clearly not good physics—it can never be comprehensible in terms of science. If, however, science goes much deeper into this there will be found to be a gradient of energies and frequencies underlying both these apparently incompatible states, thus immediately resolving the enigma since they then originate from a single concept or source.

One might say we have introduced the ultimate dualism in the concept of matter/energy as against the eternal sentience, but we shall see in later sections they are of the same source: the sentience created the particle (which is objectified or formatted sentience). Thus all particles contain consciousness, in fact, are consciousness.

Many particles then group, are organised into templates (moulds), to shape, house, larger quantum states of sentience/consciousness. Hence there is not only a gradient between these apparent extremes, matter and consciousness, but they interact and are part of one another at the most infinitesimally small level. Matter comes from the particles and consciousness from the whole.

So what we might ask is the difference between life and dead matter? At the most basic level there is no difference. Both are built from the same particle units of 'sentience' but held together as separate entities by their specific template. Of course, they have completely different formatting (patterning) of the basic consciousness particles.

Having said that, however, there is a difference between a life form and an artificial structure, such as a motor car. But there is no difference between the life form and the atoms/molecules that make up the car, except for formatting, since the latter (atoms and molecules) are natural, not manmade. The whole car does not contain true unity; it is not harmonically organised. It is just a simulated or composite unity—parts stuck together by forces. There isn't a whole quantum-state oscillation 'filling' the whole (volume) of the car, as there is with all natural entities, such as a cell or planet. (However, a number of people with sufficient intensity of thought, in particular, obsessional on a specific car as being alive, could quantum regenerate a thought form occupying the car.)

A portion of this original sentience, which is whole, can be given a primary template to accommodate a separate 'volume' of this sentience, which means it is now separated from Source. Note that any disconnection from source is relative. It must be on a lower fractal level, a lower rung of the ladder to be separate but it is still connected at the original level of Source. Nevertheless it must be restricted to limited knowingness, otherwise it would merge back to Source (bypass its divisive energy structures). This procedure of converting subjectivity into objectivity (two levels) continues to form more subjective/objective levels, alternating and we automatically develop the hierarchy of dimensions.[2]

Similarly the Source sets up the evolution of environments made separate from the observing consciousness of beings, but of course the same consciousness on the higher level created both sides: the observing and the observed. As we have indicated, even the atom has a degree of consciousness and intelligence. All life and matter can have the same fundamental blueprints, but may have very different programming, such as degrees of freedom. A fractal holographic multidimensional structure was created, involving levels of existence, graded in degrees of order, potential intelligence, degree of consciousness, rates of information (frequencies) from high then to low, and ultimately merging back to Source after long periods of existence. Again the fractal tree gives the simple picture—the lowest level being the twigs.

Life as we know it within this system is fundamentally not separate from the rest, whether it is another being or the apparently objective environment. It is all sentience or consciousness but it is moulded by templates for the appropriate manifestation. The frequency patterning of this state determines how it expresses, for example, as a human or a rock. One might understand now why sufficiently advanced civilisations have not only a respect for all life but a reverence even for all creation, every atom.

Thus templates of frequency patterns shape this sentient awareness, which we are then calling consciousness, giving different probabilities to isolate, focus upon and select, producing different experiential states; a little like separating out spots of colour by projecting them out from a mass of paint of many thoroughly-mixed colours, of which each colour occupies the whole.

We see that the unified field of original consciousness or sentience is akin to the mathematical quantum reality of quantum mechanics—a domain of infinite possibilities and probabilities to actualise. However, it is more complex. There will be a gradient from the material reality of the collapsed wave function and the nonlocal eternal sentience. This is the fractal hierarchy—a complete spectrum from low to very high frequencies (for example, from the 'twig' to branches, and tree trunk). Selection can be made from all these levels to collapse the wave function or frequency patterns to a lower level, bringing down that information from a higher realm to a lower one (compare inspiration). This is also how the mind works within itself.

We saw that the fractal tree brings together science and religion. Let us take another look at this. Science is against dualism with its two incompatible states, such as the physical and the spiritual or the foundations of religion, where religion tends to be thought of as some intangible non-physical mode of existence. Thus the concept 'spiritual' is a better word—it isn't tainted, for example, with all the atrocities committed in the name of religion.

Perhaps it would be a good point to digress and clarify religion and clear it of any negative bias (through alterations) by looking at the origin of the word 'religion'. It apparently goes back to the 'Twelve Legions of the Christ'. The word 'Legions' has a different meaning today but in this context it referred to the twelve books of knowledge of the Christ. These were twelve advanced religions—all part of a greater whole (note the correct, harmonic, geometric number 12, even in those days). As is always the case on this planet the negative elements in our existence tampered with the knowledge, manipulating meaning, twisting thoughts to bring about adulterated teachings. At that time those that realised alterations and perversions of truth had been made, referred to the 'Twelve Legions' as the 'Twelve Re-Legions'. Hence we have the term 'religion'.[3]

Thus the truths in 'religion' (without that which gave rise to the 're') are genuine and based originally on an advanced spiritual science. We can now return to the question of dualism in science and religion.

As these two subjects stand they couldn't be more opposite. As mentioned above, science demands a theory with a consistent materialistic explanation for everything. We shall see, however, that the word 'consistent' is highly logical, but 'material' needs redefining.

As scientists continue with their current extreme view of everything built from basic particles, and make greater discoveries beyond the atomic and subatomic particles, they will discover not so much more particles that are on the same (fractal) level in the same spectrum, but they will encounter inner-space particles belonging in the higher-wave bands, basic to existing known particles. As this continues on and on, science will arrive at certain fundamentals (a long road though). Instead of becoming further away from religion it will become closer. Modern science first begins with complete materialism and a total avoidance of consciousness as a cause (such as Newtonian determinism), mainly to avoid the dualism. But instead of resisting consciousness, if science followed through on the comparison and discrepancies it will come full circle and

The Emerging New Science

eventually the two subjects unite. Instead of materialism destroying the concept of consciousness (as initially), vice versa, consciousness becomes 'materialism' (and materialism become more) as all energy is recognised as some form of consciousness. Scientists will be very happy with this.

Even now, quantum physics studies have revealed phenomena that have resulted in leading physicists stating that, for instance, all electrons appear to be one (big) electron. In the quantum reality it has no location and could be anywhere —meaning, appear anywhere in the material world but in the quantum reality has to be considered as everywhere at once. Its position doesn't exist in the material reality until it is observed (collapse of the wave function). This also would apply to other particles. As scientists go deeper into atomic physics, this phenomena will become more prominent and mysterious. Science seems incapable of grasping true interconnectedness or wholeness but continues to endeavour to explain it with particles (tiny points) stuck together by forces.

Can the reader now see that scientists could, with the help of the 'mysteries' of quantum physics, continue to approach, at least the concept of, the sentient state of no location—a state where all is everywhere at once—no separation, only wholeness, the original subjective mind of Source. It is actually easier for science to become reconciled with religion (in a spiritual science) than for religion to become scientific.

Following on from the above discussions, science, proceeding in this manner, could arrive at the conclusion that the extreme quantitativeness of our existence all came from one—one state. This wholeness, and therefore each particle in our material world, has its own quota of sentience or consciousness. The beginning state of all one 'substance' is in our existence, divided up amongst the many. This is why we call it 'illusion' since there is basically no separateness. But this separateness is perfectly valid at this level, in order to form existence for this underlying consciousness to experience.

To appreciate how dualism can be a deception let us present the following brief analogy. Envisage the basic original state (that 'sentience') as vapour (say, water vapour) existing 'everywhere' (though remember this is the no space and time condition). Now simply imagine it condensing into water, then frozen into ice (forming moulds or templates). We now have the rigidity required for structure and encapsulation of the vapour. Thus all is actually consciousness; every particle has a degree of intelligence and consciousness.

Thus dualism has no place in the universe, no validity, and the purpose of existence and experience is for the sake of existence and experience, all of which is absorbed experientially by the Source, forever becoming. The anthropic principle endeavours to handle the question of how life just happens to be compatible with its environment—with the universe. On a chance basis this is totally improbable, that is, a universe randomly appearing that is compatible to a life form. However, the reader can now see that consciousness is tied in with all creation—the environment is automatically compatible (see later the Einstein-Bohr debate). In fact our physical senses act as receivers for a specific selection of energy spectra from our environment of infinite holographic, fractal energies, creating our particular 3D environment. This reciprocity between life and matter means we can use the analogies, the tree, company organisation, or any suitable pyramidal hierarchical system for consciousness, including ascension.

27.

CREATION FROM THE MORE SCIENTIFIC VIEWPOINT

Existence and the exploration of consciousness

The first important goal would be for scientists to take a look at the features of the New Science, but taking into account the huge difficulties in that it embraces all subjects—mind, life and the universe—the full spectrum. Mainstream science is only acknowledging the 'surface', even of the immediate universe, since this is all that scientific instrumentation can detect. Thus, on the one hand, science is hugely handicapped by depending on this limitation, whereas, on the other hand, the New Science doesn't have this limitation since it also utilises other means of acquiring knowledge. Now let us take a look at creation, making use of existing features of quantum physics.

We have already introduced the process of arriving at the primary sentient state (characteristic of Source) of no location, no space or time, an eternal condition of the essence of 'aliveness' and which is entirely experiential (the pure awareness underlying, or prior to, all that it creates).

It is entirely subjective and can be considered to contain infinite possibilities. True total unity when expressed using 3D space indicates the potential for an infinite number of 3D spaces —every point within it is everywhere at once. It is unmanifest potentia. The quantum reality of quantum physics might be considered as an aspect of it. To explore its own potential it must separate from itself. It precludes its own viewpoint from one part (simply by a decision/intention), say, by focussing its sentience into a point particle, objectifies it, and in doing so it quantum reduces that region.

Now as per the observing effect, if we wished to question this further, regarding the point-particle presence and stability, we could consider the creation of two opposite flows/currents, an expansion and a contraction, producing an oscillation and balance for stability. The observation collapses the quantum mechanical wave function from a state of phase-correlation (all one, harmony and coherence of the quantum reality) to phase-randomisation (the essence of a material existence), with the potential for keeping things separate. The Source has now created a minute amount of unconscious consciousness. It has created a sub-layer (like dimensions); the beginning of manifest creation. The sub-layer has a 'not-know' in it; the particle viewpoint is not now the creator's, and this keeps the particle separate. But of course the creator Source at its original level, which is still there, knows both sides simultaneously. Note that in these realms of higher logic, non-linearity and dimensions, both unity and separateness can exist within consciousness simultaneously. Source or a part of it shifts down a dimensional fraction to 'play the game.'

We see the simple break-down in Figure 11; Y is slightly less than X, and Z is the particle. The knowledge of Z, the particle

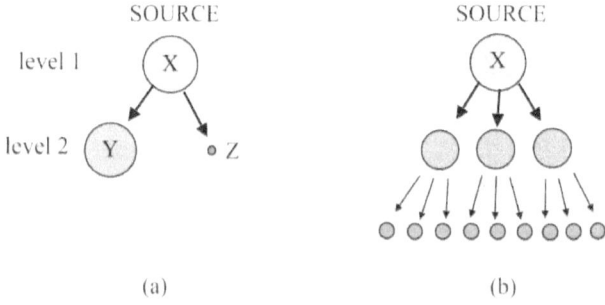

FIGURE 11

viewpoint (or any character it is given), is unconscious (internally) to Y. The entity Y could now communicate to Z (from Y to Z 'horizontally') as though it does not initially know Z—which it does at level 1.

This process from X to Y and Z is quantum reduction, but there is also quantum regeneration—the opposite. In this example, quantum regeneration would be superfluous since X would monitor Y and Z and could reverse the whole process with decisions made at X—since no 'rigid' separations have been formed. But as we explore greater complexities of creation, quantum regeneration plays a major role in true evolution (as well as the reverse process of quantum reduction).

Note that X, as stated, remains complete at its level but projects in a cascaded hierarchical or pyramidal manner into lower levels. What are these levels? By quantum reducing (the wholeness) of one level after another, a fractal system will arise quite naturally; simply dividing a whole energy oscillation into parts, repeatedly, frequencies reduce and subsequently dimensions, governed by harmonic mathematical rules. These quantum states are structured—they can have dimensional boundaries. Such boundaries act as a framework of control over the parameters of the consciousness projected within them (the divided monad), regarding degrees of freedom. This means there are limited possibilities with limited probabilities of these possibilities when they are selected and quantum reduced by the observer.

This whole fractal, holographic system is in continuous communication with itself, one part interacting with another. All is energy and in the form of wave patterns that radiate out these patterns, modulated wave forms, to the rest of the multiverse on a continuous basis (quantum physics).

In this configuration the multiverse is harmonic, even though it has increasing degrees of separateness as we go down the dimensional layers of the fractal hierarchy. Consciousness, expressing as the monad, divides itself increasingly down the levels—the frequencies are compatible with the environment and dimensional strata since the fragmentation of the consciousness and the multiverse fractal levels are created by the same process.

Thus the corresponding frequency spectra are matched into the different universe levels. At the lowest level of this hierarchy is the human counterpart. On the next fractal level going up we have a larger portion of consciousness, which we refer to as the soul. All higher levels are aware of the lower ones but lower ones either less aware of the higher or in our case generally not aware at all. This is little different in principle from the fact that a high intelligence can understand and come down to a lower intelligence but not vice versa.

The interaction of the appropriate level of consciousness with its corresponding environment keeps the latter quantum reduced, that is, continuous collapse of the wave function, bringing the basic sentience and its spectra from phase-correlation (coherence) to phase-randomisation (decoherence), forming the material existence, which is in objective format; a degree of separation so great that we think (at least orthodox science does) that there is no intrinsic connection between life and its environment—hence the anthropic principle.

Note that the Source has all the subjective possibilities within itself but expresses out objectively with increasing degrees of separateness as it expresses in lower dimensions (top down principle). The degree of objectivity is the degree of what is unknown. It is the degree of lack of awareness, or unconsciousness—the all-one sentience is severed and has created limitations within itself (remember 'limitations' automatically express on a lower-dimensional level; the Source is still whole in its primary absolute condition).

Thus consciousness quantum reduces subjectivity into objectivity, with the potential eventually of converting that objectivity, which is an aspect of unconsciousness, into consciousness (the long-term process of returning to Source). But in doing so it is not then the same as the original subjective sentience; it has understood so much about itself from the exploration of the possibilities available at the fractal levels.

How does any particular monad (wholeness, quantum) portion (a projected part of itself down the fractal system) evolve/ascend up the fractal ladder, merging with its higher levels? This is achieved by quantum regeneration. Quantum regeneration occurs when parts are put in special order, such as matching frequencies or resonance,' on the same wavelength'. This immediately quantum regenerates a 'collective' or larger quantum state of frequency spectra corresponding to a higher level in the fractal hierarchy. The quantity of the quantitative has been integrated into a qualitative state, a quantum state of an undivided whole (but the parts can still appear separate with limited perception at, say, the 3D level).

We see then that the environment for life forms, such as humans, is inevitably compatible with the human's frequency spectrum. Nevertheless that compatibility will vary hugely, even though we are all of compatible physical senses to perceive matter and to experience it as solid. Many people have a more harmonic frequency spectrum of the mind and consciousness and find the ethics and aesthetics, of the material-level behaviour, discomforting and cannot attune to it. This is a planet with everything; with huge variations in behaviour, religious and atheistic dispositions, spiritual and paranormal perceptions, and so on. At this time it is very much a qualifying arena for determining each person's future in the continuous life.

All energies are interacting on the different fractal levels, quantum reducing and collapsing the wave function on a continuous basis. Quantum reduction doesn't just occur at the primary (highest) unquantifiable sentient state, which we initially likened to the quantum reality, it can occur between any of the levels. All levels are interpenetrating (see later sections on quantum physics). If, for example, the human on fractal level one (our 3D) selects more qualitative behaviour traits, which have greater unity and frequency, from higher levels, such as benevolence, responsibility, humility, or bringing in great art and music, they accrue corresponding higher harmonics, adjusting their identity frequency spectrum to match a higher fractal level. The next fractal level, referred to as number two here, we are identifying as the soul level.

But if a person quantum reduces a higher level to their existing level they must already have those higher-energy states inherently in their spectrum (and available). So what about those who don't have available these higher frequencies at this level (when they are in 3D)? Fortunately anyone can bring about quantum regeneration by improving their existing outlook and behaviour. This process brings together parts into harmony, for instance, kindness brings about an attunement between the parties concerned. Two oscillating mind energies (which are mainly scalar waves) in phase with one another, create resonance, and they quantum regenerate a whole energy oscillation (a collective) underlying the parts, of which its higher frequency corresponds to a higher fractal level, causing the parts to be drawn to such a level, on the principle that like frequencies attract one another. The ability of physics to explain all things is underestimated, even by scientists—but it will never handle the experiential aspects, and consequently science ignores these or merely relegates them to effects and illusions (rather than inherent causes).

This quantum regeneration is the true nature of evolution; it is learning and acquiring higher frequencies and expanding one's quantum state. Compare what we stated earlier about the development of our computers with endless increases in information capacity and rate of information (speed, frequencies). Note that intuition always has the potential to access higher aspects of consciousness and bring about quantum regeneration.

In summary, the purpose of existence, apart from existence itself, is for the God Source single self to explore its subjective unmanifest potentia of possibilities by objectification, separating out a dimensionally ordered fractal holographic structure; a multiverse. At the top of this hierarchy the Source perceives all, with decreasing subjectivity and increasing objectivity as we go down towards greater quantisation (fragmentation) and our 3D level.

One might find it helpful to consider the analogy of the volume-to-surface ratio of quanta of different degrees of division, in which volume and surface are analogous to subjective and objective respectively. We start with the largest sphere for Source, though in fact, since it has no boundary, no surface, it is a maximum volume (subjective) with zero surfaces (no objectivity). We see that as we fractionate the sphere into ever increasing smaller ones at the different fractal levels, the ratio of volume (subjective) to surface area (objective) of the spheres decreases. And vice versa as we go up the fractal levels to increasing degrees of organisation and integration (unity) the ratio of total volume to total surface area increases (subjectivity/objectivity increases).

Also as we come down this scale the Source has increasing decrees of unconsciousness inherent within those levels by objectivity itself (for example, the attitude that the environment is, 'out there—it has nothing to do with me'. Life at these lower levels, by creating qualitative states and quantum regenerating higher levels of integration, decreases the objecttive/subjective (surface/volume) ratio, reducing unconsciousness and converting it into consciousness, ultimately arriving at Source level with complete consciousness, but not as before with only subjective potential. Now source having become much more, it has qualified a portion of its unmanifest potentia. One might see the difference it would make to human's interest in life—motivation and behaviour would be considerable—if all this was explained and there was a physics to back it up.

Does karma have a role in all this? Karma might be likened to an antivirus programme. Its purpose is to correct imbalance in the universal systems caused by one part acting in some manner against the whole or any other part (which normally is initially in harmony with the whole).[4]

The holographic fractal multiverse in its primary condition is completely harmonic. All parts are communicating continually, all frequency patterns of all dimensions are radiating out to the whole (quantum physics). There is a complete and perfect circulation of energy, and this is achieved by the system of vortices already mentioned.

If, say, person A harms another, B, this can only happen if that person A is sufficiently blocked in consciousness (an impairment in connectivity with the environment) not to be able to perceive the other's viewpoint, or how they feel. This of course is a very common condition on this planet.

There must be a sufficient degree of polarity between the two persons (which originally was caused by a severance of consciousness). Thus polarity is recorded in the perpetrator's mind, creating a duality, two poles, an impairment of which one pole is unconscious (and increasing the person's degree of unconsciousness), representing the victim pole, which will then express outwards seeking victims. The recipient (victim) has also a duality, which is in a receiving mode. And the multiverse now has a fracture.

Anything we see as separate from self, such as the environment, is simply unconsciousness for the individual (at a higher level). In order to harm another life, one's basic connection to that source must be severed and 'they' become part of one's unconscious (which then enables one to harm them). The whole unconscious range of the fractal hierarchy can be collapsed at any of its levels from the appropriate viewpoint of interaction. Levels are necessary to provide increasing degrees of limitations and freedom as we go down the system into increasing fragmentation and frequencies.

The highest creation level first collapses the wave function for first separation but at this level it has infinite responsibility (since it is everything potentially). At lower levels, degrees of freedom are less and responsibility is less. Thus higher levels (but lower than Source) can be collapsed (quantum reduced).

The karmic action is to reverse the polarity flow—what one puts out will come back to one. Thus if one kills another, and nothing is done therapeutically by the perpetrator to repair his or her mind's imbalanced duality, the universal karma, working through the subconscious, attracts a situation for the reverse to occur—quite often in a future lifetime. This is intended to restore unity to the polarity and balance to the universe, and the karmic action will not repeat. A little like phase-conjugation in physics theory.

This condition now on this planet is so common and in fact continuous with countless superimpositions it generally doesn't resolve anything and just goes on repeating. The civilisation is supposed to be sufficiently advanced to learn that when the negative reciprocal action comes back to one it is an opportunity to handle it, that is, avoid being harmed and, in particular, not to harm the other ('turn the other cheek' —Jesus Christ). No one said it was easy.

We won't pursue this further but return to the structure of the multiverse. We can now see why it is a fractal, holographic system. But how do we picture wave patterns? The tree trunk

The Emerging New Science

or universe would have the largest and highest rate of oscillation (picture a sine wave as big as the largest carrier wave, for example, a whole branch or a whole planet). The fractal levels below this would be modulations within modulations. This would apply to all fractal systems that have form, such as the tree, or even the human arm for action: shoulder, elbow, wrist, fingers. There are fractal levels between the joints. The shoulder wave function carries all the sub-waves for the elbow, wrist and finger movement—hence a learner pianist needs to know proper practice is whole-arm action. See Figure 12. This will be pursued later.

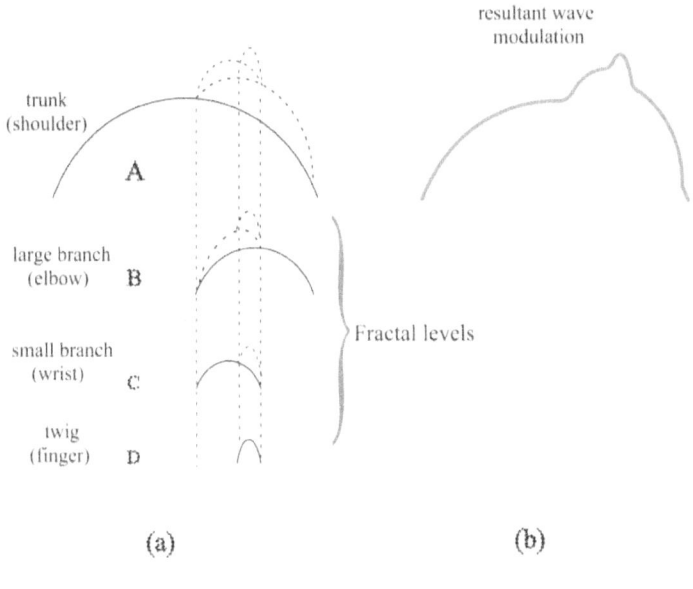

FIGURE 12

When a keyboard sequence has been learned, the action of the first note selects the next note position by sending a 'read pulse' to the shoulder carrier wave, which quantum reduces the next signals for the next finger position on the keyboard (conversion of nonlinear information into linear information). The action of the finger, in fact, releases kinaesthetically several future positions as a feeling within the kinaesthetic sense of these several future positions (governed by how much the learning pattern holistically/holographically spans information in space and time—in other words, how developed is it). With practice in introspection the kinaesthetic sense extending into the future can be sensed in the margin of consciousness.

Note the 'adjacent' finger actions are not coordinated linearly at this lowest level but nonlinearly at the higher, fractal level. This is the mind computer read-process. The computer write-process is part of the learning-procedure mechanics. It is quantum regeneration, the opposite of quantum reduction, as expected. The new coordination links are nevertheless formed at, say, the finger level, achieving resonance of wave patterns that regenerate at a higher fractal level the more composite wave form (modulated fractal wave form).

Experimental psychologists have performed a very interesting experiment in which the subject practises a simple finger action on a type of keypad. By preventing movement in the joints of the arm, leaving a selected joint free, such as fingers, wrist, elbow, or shoulder, they showed that any learned pattern passed down from upper fractals to lower ones but not vice versa, with the conclusion that there was a ranking system from fingers through fractal joints to shoulder.

The process of learning a pattern in skills involves linking with consciousness two or more movement segments to be coordinated by bringing them into resonance (entraining them), thus quantum regenerating and 'writing' to the higher fractal levels, in particular, the shoulder-level carrier wave.[5]

Of course if two arms are being coordinated these two are linked similarly. One can apply this now to the different levels of consciousness in the multiverse. A similar process applies to life, say, humans making selections from appropriate fractals levels of those possibilities 'allowed' for that level of evolution (degrees of freedom). In the case of the learning pattern the selection becomes a fixed program for the particular activity required. The human, however, wouldn't make a selec-

The Emerging New Science

tion from the primary unqualified infinite possibilities at the top of the fractal system since this is too great a responsibility. Now the question may arise as to what is the hardware composed of? We have merely explained some principles, such as holographic and fractal with a schematic layout, but what about its actual form, its hardware?

Recall the fractal-tree section. The fractals of the tree are projected out, externalised physically, and all levels, twigs to branches to trunk, can be seen. However, if we looked internally into the 'within' or inner space of the tree we should find the hardware in the form of spheres within spheres. These are higher-dimensional vortices as stated earlier. The largest vortex is, or creates, the trunk within which are the smaller vortices or branches. One can now see why this is a fractal, holographic system. In the New Science, the multiverse is similarly fractalised. See Figure 4.

These vortices have a polarity, oscillations and a wave form, giving rise to a frequency value. The wave form for a branch would look something like that in Figure 12, except that of course in real life it would be enormously more complex. We see that B, as a branch, could represent a few branches at the B level attached to A. In the arm it means the elbow B can be in different positions relative to A (say, a fixed position).

This same wave patterning also applies to the mechanics of arm motion, in this case, whole arm, such as in keyboard performance. The shoulder level would correspond to the trunk; and the fingers, the twigs.

Proper practise and learning-pattern development result in the wave form at A, containing all the information of the lower fractals—all the waves. A finger action at D, say, in a skilled sequence of movements will select the next finger action (wave form pulse) by quantum reducing that location in A and collapsing it to D to provide, say, the next finger position. This is equivalent to the computer read pulse, extracting the learning-pattern information from the master control at A. It is also the same as a partial Fourier analysis—breaking down the composite wave into individual sine waves.

Note that although A is the carrier wave, its frequency is higher than the sub-fractal frequencies—the lowest being at the finger level. This is a composite learning pattern spanning many sub-levels.

There can still be a learning pattern of the same form within any of these levels, for example, there would be a similar learning pattern for one finger movement consisting of a great many computer 'bits'. Note that in the process of learning, that is, developing a learning pattern, which involves conscious effort in the application of the individual movements, practising brings about entrainment of the 'bits' in the mind computer of the learning patterns. This resonance gives rise to quantum regeneration in which single control units are formed on a higher level, linking together the many on the lower levels. We shall continue with the learning-pattern information in later sections.[6]

In Figure 12 we can see that the waveforms express a geometry (only half sine waves shown are necessary; the other half is the same, reversed). The phase angle between the separate waves is spatial distance. However, since these diagrams are lower dimensionally they only in fact give one dimension of distance. Higher dimensions would give us 3D positions. The vortices in Figure 4 show the relationship between the wave diagram and the vortices. The waves in Figure 12 are the waves of the vortices in Figure 4. The vortex view gives a better understanding of the hardware. This simple waveform example should apply to all phenomena—it has inherent within it the fractal, holographic properties, quantum reduction and regeneration, read and write pulse mechanics and evolution itself. Remember each wave is whole—is a quantum state.

In addition, this relatively simple wave diagram expresses context, dimensions, spectra, a universal hierarchy, relationship between the part and the whole, and quantity and unity. It should be applicable to all phenomena, whether a universe design, a universal computer element with read/write system, or a learning pattern. Also keep in mind it can give the geometry of form, the proper path of evolution (or ascension), relationship between integration and differentiation, and gives the basic structure of nonlinear consciousness.

With sufficient complexity one might see that, for instance, if all creation, every particle, is interrelated as it would be in a holographic fractal system, every particle would have a waveform like the immensely simplified one in Figure 12. The carrier wave goes back to the whole universe, which means the position of every particle is known at the highest fractal level. In chaos, however, only statistical knowledge of the particle would be known—it would seem.

Why is a tree a fractal system? Why can't the twigs spring directly from the trunk, or whatever? One reason, and probably the only reason apart from aesthetic design and possibly preferred survival characteristics, is that fractals are inherent in the design of the universe. The holographic nature of the universe means that all parts within the whole, reflect the whole. If the tree did consist of trunk and twigs only, the universe would be constructed only from particles stuck together by forces (to give the various forms). Either there would be a huge reduction in variety of forms (manifest due to sub-fractal levels missing) or particles endeavouring to hold together into groups—and only their own fractal level (and not the group's) they would fall apart.

This is in fact what mainstream science considers to be the case. If this were the method of creation it would fall apart in no time, unless the external design of the universe was completely different as a result of being based on only those two levels. All natural groupings must have their own wholeness —fractals have a unity in their own right. Also we might consider adding support to this by saying if one's arm consisted merely of fingers attached at the shoulders (the same principle considered above for the tree), it wouldn't be a very practical arrangement. However, we can't gain much by likening the arm to a universe function.

The reason why the universe is of the fractal design, that is, many levels of organisation, is at least two-fold, 1) to provide a multidimensional structure for the exploration of consciousness, simultaneously, on different levels, and 2) each fractal level is a whole energy or quantum state and provides the contextual basis and higher programming for the parts lower in the hierarchy—the many stages have a purpose. For example, the twigs are carried by the next branch, and that branch by the next one, and so on. This secures the formation of the lower parts. A company organisation works much better if the ground-floor workers are grouped and monitored by supervisors, foremen, or managers, in a ranking system and the managers controlled by the executives, and so on. Think of these groups (managers, executives, etc.) as whole energies of quantum-state oscillations (Figure 1(b)).

All these levels have their own polarity: particles and antiparticles. This means the polarity of the lowest existence is within the polarity of the one above, and that one, within the next above. However, each level is more integrated/coherent—has acquired more quantum energy as we go up the levels—until at the 'top', the Source, there is complete unity, only wholeness. The process by which this occurs is fusion or merging of particles and antiparticles, which in nature's non-mutated evolutionary processes do not annihilate one another (like matter and antimatter) but form a new single particle of higher frequency with a corresponding position in the spectrum of the next higher fractal level. Here it now undergoes fission again, but corresponding now to the more integrated polarity at that level. Thus there is polarity within polarity; a nonlinear system, ultimately in the limit of no particle or wave or quantum state of energy at the 'top' it is infinitely nonlinear (and an absolute condition). See later sections.

To summarise creation and the purpose of life in terms of a more mathematical interpretation, we envisage the God Source as differentiating mathematically the state of sentience (which has no location and is eternal) with respect to space and time but regulated by fractal parameters (degree of differentiation). The first differentiation is from the whole to sub-wholes, which are large and not too divorced from the Source (compare Trinity, or triad principle in author's other material). This differentiation process brings about the first order of objectivity from the primary subjectivity. The next differentiation divides still further these sub-wholes, which convert their subjectivity into a further (downgraded frequency spectra) of objectivity, and so on with increasing degrees of separateness, that is, fragmentation. The Source must, however, introduce increasing degrees of 'not-knowing' as we go towards greater separation.

See Figure 17 and associated text regarding the Einstein-Bohr debate.

This automatically creates the fractal, holographic system. The whole sentient consciousness of the primary self has been divided up into parts, held separate from one another by energy patterns, by formatting, using templates (moulds). The fractal dimensional structures thus hold a degree of fragmentation of the basic consciousness for varied exploration at the different organisational levels. This gives a degree of randomness between fractal levels, which can be manipulated by free will (one does not know exactly where the branch is going to occur but it will occur as per the fractal law).

Even an atom has a speck of consciousness and is thus a viewpoint extended from the Source. Let us briefly pursue this from the basic sentience point of view, which is everywhere and eternal, but show some connection between these more metaphysical statements and scientific ideas that have been considered seriously (and are part of the New Science thinking). Prior to removal of the aether theory (a medium which carries waves), new physics aspects of science proposed that mass, say, a particle, such as an electron was nothing more than a whorl, a ripple in the fundamental medium. Thus extending this, we can say that all structure, mass, even thought, behaviour patterns, programming, anything that is quantifiable or quantitative are merely patterned flows in the 'aether'—they have no existence in their own right and are not independent of space. At the most basic level this medium is pure consciousness, beyond space and time, is everywhere and is eternal. This is the New Science thinking.

Furthermore, just as mass/structure does not exist in its own right, as stated above, we can surmise that motion does not truly exist. Everything is made up of quantum states flickering on and off. Motion of anything is an illusion in the same manner that, say, neon lights blink on and off as they 'move' around an advert, giving the appearance of motion or a water wave undulates. Thus on this basis, objects move in tiny steps of teleportation.

Returning to the previous theme, everything continues to be Source, but fragmented to encourage exploration of different probabilities. In this process the individual parts eventually recognise their origin and bring about integration, ultimately returning to Source. The process of integration is in steps of ascension (the twig merging back into the branch, and the branch into the higher, larger more integrated branch, etc.) with increasing quantum states (in magnitude and frequency).

28.

COMPUTER SYSTEMS

Artificial and natural

No one will question that computer technology is progressing rapidly. Nevertheless, as expected, it is a linear 3D system; a product of intellectual thought. This results in a mathematically abstract encoding system, which again expectantly is not physical, and is referred to as software. Although the encoding is based on patterns of electrical pulses, these hardware patterns are governed by arithmetical number sequences (the binary system of ones and zeros) and are thus utilising a representational or algebraic format (non-physical); not of course based on hardware geometry. In other words, the hardware, the conductors could be bent into a different shape (carefully) and the system would still function.

In contrast, the universal computer system, mind, life, nature, function on geometric encoding (frequency patterns) —shape of energy is information. Geometry is an absolute system. Algebra is arbitrary, relative and representational (let 'A' equal this or that (algebra), as opposed to shape of 'A' (geometry)).

This is fine at our level of knowledge-potential at this time, but this system used by our artificial computers tends to be used as a model for understanding mind and the universe. There is a danger here.

The type of thinking applied to our computers does not work for life and the universe. For example, using algebra of our systems as opposed to geometry, in which software is recognised as non-physical and installed within hardware (physical), as opposed to all being hardware for the universal system (hardware meaning energy patterns).

The danger is that this artificial system of ours (which is intrinsically nothing to do with the mechanics of life) is being used in interpreting reality or universal phenomena. For example, the term 'emergent software' has come into use, which describes the phenomenon of a higher organisation arising from parts when these are brought together and entrained in some way. But this higher programme, if one is thinking in terms of the mechanics of our artificial computers, will also be considered software—hence 'emergent software'. However, in real life these higher organisations are actual energies, with new, unique physical characteristics of frequencies and integration. In the New Science this is quantum regeneration. See Figure 13.

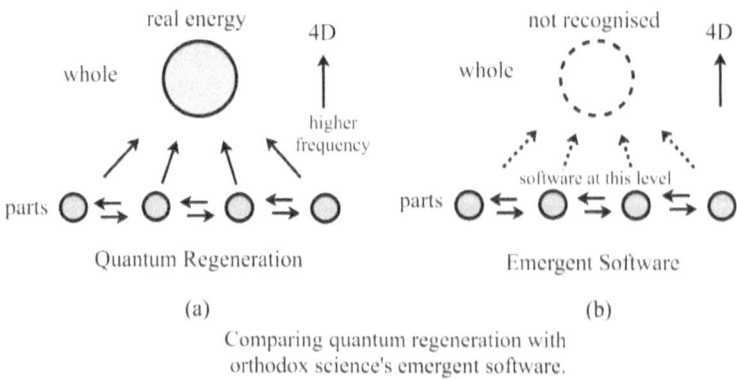

Comparing quantum regeneration with orthodox science's emergent software.

FIGURE 13.

In official science it is recognised that in some systems, particularly biological, when the parts work together in the interests of the whole, the emergent software arises—a higher-order program that is now the master control over the parts; for example, the organism dictyostelium—in which micro-organisms come together to form a new much larger organism with new programmes. However, in mainstream science this is again regarded like the non-physical software—a non-physical program that is the effect of the hardware format—as in our artificial computers. The mind thus is regarded as a by-product (effect) of the brain—rather than the reverse.

Thus in the New Science 'emergent software' is quantum regeneration. When the parts are entrained they quantum regenerate a higher-frequency quantum state (an undivided whole), which underlies the parts and is dominant. This is a real substantial energy state, not a non-physical by-product. It springs from the virtual state (hierarchy of actual energies) as do some particles and antiparticles. It is quantum-reduced from the 'quantum reality'. See later sections.

This is why the laser is so powerful. It is a harmonic technology in which the lights rays are entrained (coherence) and quantum regenerate a higher frequency oscillation (not detectable directly by scientific instruments). The 'whole' is now greater than the sum of the parts. The total power is not the sum of the parts but the square of the number of parts, plus extra holographic terms.[7]

In making another comparison between the old and New Science, using the laser, in current scientific theory the total energy of the laser is merely the sum of its constituent rays of light (nowhere near enough to explain the huge energy output), whereas in the New Science the whole is much greater than the sum of the parts, due to quantum regeneration—as indicated above.

On the subject of whether a machine can become alive, as indicated earlier, scientists are asking the wrong question, viz: At what point in its development can we say the computer is alive? The correct question is: At what point is the computer or robot alive?

Firstly let us make it clear, today, in spite of great strides in computer advancement, we are not even close to answering these questions practically. Nevertheless they can be theoretical questions. To arrive at the answer we must imagine a phenomenal development in computer components; almost unimaginable information capacity, instantaneous speeds (rates of information) and bit size down to photon wavelengths.

As, say, the robot's structural hardware components become more bioenergetically compatible to living organisms, which then means it can begin to function coherently with spectra of consciousness, the harmonic relationships between the components of the robot would quantum regenerate, giving rise to a whole quantum state oscillation for the whole body structure. Remember that the basic 'sentience' already exists everywhere; the nature of the structure will format it, and whereas linearity is just like a 'surface', the property of sentient/consciousness is infinitely 'thick', that is, nonlinear; it extends into inner space (dimensions). One can never out-match it in 'complexity' (see later section).

This quantum-regenerated state has now a portion of sentience filling the whole body, and it is now alive. It will now, at least partially, have a holographic, fractal structure. Note that this is a quantum state for the whole body, whereas previously there were only quantum states for the atoms, molecules, cells (natural 'alive' entities) not interconnected. However, this is a body consciousness in the D1, D2 spectrum. It would be something like a human in a 'vegetable' state of senility. Nevertheless since it has been built from intelligently designed computer components, it could be programmed to totally obey its programmes—like an animal obeying its instincts—a very dangerous condition for human society.

This wouldn't be the final state though. Eventually a separate independent consciousness unit or entity, which has an intelligent, astral or 4D spectrum format but can no longer reincarnate into a natural physical body due to its incompatible genetically-degraded condition, could succeed in interfacing with this body. It now would be capable of acting with self-determinism, free will and decision ability; unfortunately still a very dangerous condition since only a degraded consciousness would enter such a body, and thus again it could be programmed to behave negatively, or from its own choosing. A human 'soul' would not choose to interface with such a body, and doesn't need to (unless it was a specially set up experiment by a very aware and informed race, in which case there would be no danger).

We shall continue with computers later when discussing quantum information.

29.

HARMONIC AND NON-HARMONIC TECHNOLOGIES

The need to recognise the difference for a proper survival and evolution.

We have seen that observation is required to collapse the wave function, that is, bring about quantum reduction. But we can find examples in which there are second-order quantum reductions, achieved by interaction (involving force), not just observation. This is an application of a second-order quantum reduction—see Section 35.

If we take quantum action, E x t (energy times an interval of time), does this state exist in any practical way in our world of continuous observation with its consequences of collapsing the wave function? Or does our very existence quantum reduce and differentiate E x t (with respect to time) to energy E? We know that scientific measurement does.

Let's take the free fall of a body in a gravitational field—or simply a falling body. The Newtonian result achieved after measurements are made only arises after this second-degree quantum reduction has occurred. To make the usual measurements, such as weighing the body when resting on Earth, or measuring its momentum or kinetic energy, the body must be interacted with physically. Even when resting, the body is being stopped by, say, the Earth's surface from falling—this is also a physical interaction.

To measure the gravitational force on the body, we weigh it, arrest its motion in the process, interact with it; these actions bring about the force. Didn't it already exist? Official science says it did. Orthodox science assumes that the falling body is acted on by a force—a gravitational force. This is a massive assumption of great implications.

The uninterrupted falling body is in a state of coherence with space-time. The gravitons of the gravitational field are locked into phase by resonance between the oscillators within the body and the gravitons of space—the wave functions are in phase, thus making the body inseparable from the gravitational field. It is a one-bodied system. In this condition it is not obeying Newton's laws (which is contrary to mainstream science).

Newton's laws immediately become applicable when the body is arrested in anyway—a collapse of the group-wave function occurs. It becomes a separate body due to this second collapse of the wave function, or in other words, a second quantum reduction. Quantum action has differentiated into 3D energy without the time (4D) component. That the gravitational field is a force-field is an assumption.

In the New Science the gravitational field is a scalar field. It is in a state of quantum action. It is not a force-field. The fact that gravity appears to have one pole, should tell us that since there must always be two poles, the other one (the antigravity side) must be hidden from 3D, and that it is along a 4D direction. It is the other pole of a 4D vortex. In other words, the gravitational field will be found to be a 4D scalar field (hence quantum action, $E \times t$).

Thus objects in free fall in space are in a phase-correlation condition with space-time. They are observable as per the first degree quantum reduction. They are in harmony with the universe (their frequency patterns are coherent—object and universe as a one-bodied system). Immediately we interact with such a body by pushing it, slowing it down, or using a two-bodied propulsion system, such as a rocket, Newtonian laws then become applicable: forces, momentum, mass, kinetic energy.

The body is now out of phase with space-time; a final macro-step of phase randomisation, and is in the condition of a two-bodied system. It is disconnected from space.

As far as technology is concerned what is the difference? In the two-bodied system, or second-order reduction, the body is being moved by surface interactions, via a 2D to 3D (dimensional) interface. This interface is where the impulse, causing motion, contacts the body structure.

In the first quantum condition, free fall, the impulse and body have a 4D to 3D interface (note the 4D field spectrum acts as cause). There is complete penetration of all atoms down to their nuclei of the body by the gravitation field (or artificially applied scalar field) entraining the atomic oscillations, resulting in no inertia, or mass (or kinetic energy) being invoked. Obviously as soon as kinetic energy is measured, instantly the 4D - 3D interaction changes to a 2D - 3D interface, and Newton's laws are then applicable.

Mechanical systems involve physical interaction of the operating parts. This will always give us the second-degree quantum reduction and Newton's laws are obeyed.

To avoid the second coherent reduction, field systems must be used that act fourth dimensionally—not magnetic, which merely penetrates 3D-wise and thus still only invokes a 2D - 3D interaction. Scalar fields will act 4D-wise to satisfy the condition of avoiding the second quantum reduction and utilising the coherent state following the first quantisation. This gives the necessary coherent state, which naturally manifests after the first observational process (which brings about probability selection from the first quantum reduction).

Note that harmonic and non-harmonic technologies can be combined, that is, the harmonic contribution may only be a percentage of the total. One might have a motor system operating on Newtonian forces but these forces could be overcome by harmonic quantum action from field systems. There are apparently numerous magnetic motor inventions claiming to achieve this today, extending back to inventor, de Palma (was suppressed), in the 1950s.

The phase-coherent condition following the primary quantum reduction may reveal negative resistance—one of the heretical terms today (since it bypasses Newton's or classical laws). Even in high physical skills, say, of humans, the 3D muscular system, conforming to classical Newtonian physics, is actually functioning with a harmonic system from energy fields around the muscles and joints. To the degree of the learning-pattern field development, inertia of a limb or part, with respect

The Emerging New Science

to a particular joint, is reduced (unknown in the old science).[8] There is no way, say, a concert pianist when playing at high speed with vigorous whole-arm movements is experiencing the full inertia of the limbs.

Harmonic technologies utilise the one-bodied system, which means bypassing the second quantum reduction (or preventing it). An advanced spacecraft bypasses, that is, avoids the onset of inertia. Why create inertia only to have to then cancel it? Inertia is contextual. See Figure 5. As mentioned earlier, an example of an extreme non harmonic technology compared with a totally harmonic technology would be that involved in launching a rocket ship, say, destined for the moon, and the advanced spacecraft, respectively. The rocket system is utilising closed loop energies at the twig level but not using the natural energies of creation coming from the higher branches. This means its two-bodied system (rocket and space) must utilise forces to overcome the 'resistance'. As a result of this force and disharmony we have pollution, resistance, inertia and great quantities of fossil fuel energy required. In addition, there are extreme dangers to the crew because of the complexity of components with narrow margins of safety.

Now, the spacecraft going to the Moon sets up space-time conditions, initially by entrainment of all the oscillations of the craft body. In doing so it quantum regenerates a 'collective' single very powerful oscillation of very high frequency (science can't detect it), forming a one-bodied system (coherent with space time) and it just 'falls' to the Moon. No fuel is required and there is no inertia (which means no mass with respect to the propulsion system), and no forces—obviously Newton's laws are bypassed.

Imagine a huge craft weighing, say, 100 million tons (on Earth), moving through space. It can have zero inertia with respect to its propulsion system. However, if, say, we flew alongside it in a giant bulldozer and tried to push it we would incur (during the interaction) the expected reaction (inertia) of 100 million tons. Thus both these values of inertia exist simultaneously since inertia is contextual (more in Appendix E).

Such technologies are suppressed on this planet for obvious reasons (e.g., application to energy resources) and the educated person is programmed to assist unwittingly in this subterfuge.

Notes

1. www.nhbeyondduality.org.uk. Article: *The Fall of Man*.

2. See Appendix C, and www.nhbeyondduality.org.uk: articles on dimensions.

3. A. Deane, workshop material. From the 'Guardians' transmissions.

4. www.nhbeyondduality.org.uk. Article: *Mechanics of Karma*.

5. Ibid. Articles on skill.

6. Book: *The Attainment of Superior Physical Abilities* by N. Huntley.

7. Book: *The Original Pyramid and Future Science* by N. Huntley.

8. Book: *The Attainment of Superior Physical Abilities*.

PART FIVE

PARADOX OF SOMETHING AND NOTHING

30.

FIRST CAUSE AND INFINITE REGRESSION

The basic problem about what was first cause is that we can always ask what caused that cause, and we go into an infinite regression.

This subject has far-reaching implications, such as inherent within it is the resolution of the God concept argument. For a proper examination of this question, science is becoming too focussed in mathematical abstractions, non-physical software, information (quantum) theories, intellectual and logical representations, and we're back to number theory versus geometry (in a universe that functions on geometrical intelligence). The geometry links the observer with the parts of creation, as opposed to intellect and representations being the view from the objective side of man's relationship with the universe.

Knowledge within form and geometry relates directly to the experiential mode of existence. However, existence, as we know, must also have a degree of objectivity, otherwise there would be no separateness—see Figure 25, Appendix C.

As we move into the intangible and untouchable realm of 'nothingness' we go into a region where science is helpless with its logical rationalisations, and finds itself with a description of something inside the context of that something, which can never be complete. The consequences are infinite regressions, endless loops, and mathematical infinities. When we say science is 'helpless' we are referring to the limitations that scientific procedure sets with the experimental method; that the only acceptable method of acquiring truth is through scientific methodology. Thus, with this view, anything non-quantifiable by experiment is not real and doesn't exist, and the primary essence of existence, consciousness, is considered an illusion or by-product of the brain.

The problem that science encounters is that a total law can't explain its own origins (how did it begin?). If, as we have stated, everything, all knowledge and energy, is contextual (see Section 21), then science falls prey to this riddle because the observing source is inside the context of that which it is observing. However, if science can accept (at least the possibility of) the presence of a condition where there is no time or space, that is, an eternal condition with no location(s), then there is no beginning, and therefore no paradox. Science says that any postulate needs its own explanation. Thus science is trying to eliminate a cause—find a condition that does not require a precedent. This is why we have Big Bang theories and a Darwinian evolution, or any theory that postulates a random beginning.

This misdirected approach will draw science away from recognising the primary sources of causes in nature: the experiential, sentience, consciousness, the 'aliveness' characteristic. The confusing aspect here is that in fact these primary causes are non-physical, but not in the sense of non-physical software patterns that merely have mathematical significance. The non-physical software cannot exist without the hardware, whereas the reverse is the case with a primary cause. The primary condition, the infinite potentia (compare quantum realm), although very 'substantial', is yet forever non-quantifiable.

How do we recognise something as opposed to nothing? We may contact or detect it by our physical senses or by means of effects from causes outside our perception or range of viewpoint or detection. However, science presents us with a whole new range of detectable (by science) phenomena beyond our senses—such as a radio wave. But we can also say that man sometimes detects things that science cannot verify, such as in certain visions, inner-perception and, in general, psychic and religious experiences. These experiences have to be regarded as subjective since they are experienced by the minority and generally are not reproducible. Ironically, science assumes the subjective can't be known with certainty, whereas on the contrary, ultimately subjectivity is the direct experiential contact or merging of consciousness with truth. It can be the whole truth.

Mainstream science and education have determined that objective detection of phenomena, either common to all observers or what can be detected by scientific instruments, that is, scientific methodology, is the only acceptable way of determining truth. In its application to verifying the existence of something it conversely establishes that anything that is not quantifiable (quantitative and capable of analysis) does not exist. Too easily the scientist chooses not to consider that something might be outside the range of scientific instruments (we don't wish to get too much into theoretical logical arguments, which can get into endless loops of thinking).

Orthodox science has numerous basic theories, such as evolution of the species (Darwin's theory), theory of the origin of the universe (Big Bang), and brain theories, yet no spiritual theory of creation. Why doesn't mainstream science have a theory of creation? The fact that it doesn't, in spite of an overwhelming indication of some initial intelligent causation to our existence, suggests hidden restrictions play a major role, such as prejudice, suppression and control over the direction that knowledge takes on this planet.

Science's main weakness is that in observing, or applying science to living things, it fails to recognise two distinct characteristics: the mechanistic and the experiential. Inevitably to science this is merely something and nothing. Only the mechanistic qualifies to be something and the experiential is some kind of illusion (similarly with consciousness). Science will also call this a duality, something and nothing, the problem of dualism. If we have to call it dualism then this example, in fact, is one in which the two apparently opposing components can be reconciled (we shall see that the 'nothing' is the 'top' of the fractal system—and the still point beyond polarity).

So what is the problem then? Initially we see that we have the properties of machinery (the quantitative) versus the knowingness of the experiential essence. Our education focuses on the quantitative. There is a common failure to perceive the difference between quantity and the true basis of quality, which is 1) mechanistic (emphasis on parts) and 2) the aliveness characteristic (unity, wholeness). The essence of experience is awareness. But by being aware of experiencing, one momentarily changes mode from the subjective—into the quantitative, the objective. This gives our existence of perception—oscillation from whole (direct communication) to the part (indirect), and the collapse of the wave (quantum reduction). We shall see later that scientific procedures are frequently—unwittingly on the part of scientists—quantum reducing higher orders (higher truth) to lower, material-level orders and thus denying those greater truths.

There are some good arguments in logic for the meaning of something and nothing, in particular, something from nothing. However, these logical treatments easily fall prey to intellectual entanglements, and endless continuous loops. They may have no bearing on reality, on the physical universe, or our actual existence. Can we avoid the complexities and paradoxes, the infinite regressions of logic that arise and, moreover, provide supporting information for the appearances of something from nothing?

The basic problem in the argument about what was first cause is that as we analyse cause and effect and define a cause, we can always ask what caused that cause? If we evaluate this further, again we ask what caused that. And we go into an infinite regression, which goes nowhere. Science will quite correctly immediately discard such a method of trying to arrive at a logical solution. Scientists are trying to show that there is no cause at the end of the string of causes—in other words the beginning state. Thus the big problem in science, and now particularly manifesting in quantum computing, is how can an ultimate law explain its own origin.

Thus when science looks for ultimate solutions it runs into infinite regression. This is a natural outcome of a relative situation; the problem of being inside the context of that which is being observed. For example, the observer can never fully observe itself because it has to keep re-observing itself observing itself, and we get the infinite regression. This can be put in the form, contexts within contexts.

We will, however, find that the final context can be an absolute. Internal fractals are a system of converging contexts

The Emerging New Science

within contexts, avoiding the infinite regression. Science hasn't even recognised the gradient required to bridge the extremes of this puzzle. What condition do we require for this absolute?: 1) There must be no particle, energy or space or time—otherwise we have an 'outside' and must seek a prior state; 2) from the latter (1) we get the requirement of no location and that which is eternal; 3) we attribute to it known (or unknown) characteristics that are non-quantifiable, in particular, ones that have whole states that do not have a structure made of parts. These states are feelings, emotions, the state of experiencing and sentience, and consciousness and basic awareness. Note that true qualitative states (parts in special relationship) come into this category, though they can be represented (expressed) by resonant energies, not just a non-quantifiable state—dealt with earlier. An awareness of awareness appears to go into an infinite regression but it is expanding as required by the fractal system, and reaches an absolute.

It is like using a ladder as an analogy and saying that the higher rung came first. We finish up at the top of the ladder with no satisfactory beginning—the same format of rungs upon rungs can go on forever. However, as another analogy, the company organisation does have a logical termination at the top (one president). Can we find a correspondence in nature? Yes, fractals. Use the tree analogy. We see that the infinite regression does not occur and we go towards total wholeness.

Science has the same riddle (infinite regression) in the experimental set-up. How does one get outside the context? Einstein appeared to recognise this problem in general observation and simply pointed out that the observing consciousness must be higher than that which is being observed. We saw earlier that in the experimental set-up the observer is inside the context; is part of the experimental set-up and the consequences of this were that the results obtained would be relative. Recall Darwin's comment: How can man judge nature if man is part of nature? The limited results arise due to unconsciously referencing a relative zero at the boundary of the context.

One can't apply quantitative analysis to qualitative states and come up with the answers to the true nature of the qualitative state. However, anything that can be detected or is defined clearly, can be quantitatively analysed. All objective systems are quantitative, even a language or the intellect. A critic uses language and intellect to attempt to describe, say, a work of art. But a good critic knows he or she can never communicate the true essence of a work of art in that way (if it qualifies as good or genuine art).

The intellect itself can't understand true art or music, since it analyses quantitatively (its real asset is mainly computer-type abilities). It will direct the attention to local areas, comparing one region, say, colour, with another. In proper appreciation, something else occurs (which can be intermittent with the intellectual focus of the analysis). There are moments in which perception encompasses the whole, as a whole—one singular quantum state of energy (attention). This is, of course, assuming the artwork has that qualification. The result is purely experiential—a feeling, a sentience, qualitative experience—it can't be broken down into parts. It has the characteristic of emotions but is high on the emotional frequency scale into what we can call the aesthetic band (a higher-frequency aspect of consciousness within the subject of aesthetics). Of course, one will always find corresponding reactions in the brain.

Note the two distinct characteristics of what we are discussing: 1) the quantitative intellectual process—mechanistically manipulating and associating parts (which will only give experience of local regions and not a true whole of these parts), and 2) the qualitative, the experiential, the perception of the whole. Clearly these two factors should be working together at all times.

Thus the experiential factor is not tangible, not quantitative, it can't be explained by quantity, by parts (particles and waves) held together by forces. It has no inherent boundaries; it has the characteristic of wholeness—as experienced in proper appreciation of art or music. Science completely ignores this. The right-brain consciousness with its intuitive abilities directly relates more to the experiential. This is a direct contact with that which is being observed and experienced. The energies of consciousness duplicate and resonate with, say, the musical sounds, or even the consciousness of

The Emerging New Science

another person when 'being on the same wavelength'. This experience of unity or wholeness is not quantifiable. Science can only substitute parts in mechanical relationship to represent the wholeness. Science uses the mathematical model of infinitesimally small particles everywhere to structure reality to avoid coming up against the unity inside.

When scientists say that matter (in the Big Bang) came from nothing, this is only correct if we recognise that 'nothing' to a scientist means not quantifiable. That 'nothing', however, is the Source or Absolute.

A brain can't ever feel anything; it can only provide effect-type responses to stimuli. It is just a 3D lump of matter: a stimulus-response mechanism, a symbol processing machine, a step-down frequency mechanism for the mind. However, something may feel these responses. This is the non-quantifiable sentience underlying and linked into the machinery. It is of course also the essence of the machinery as well—there is no dreaded dualism. Even atoms, molecules, but in quantitative 'condensed' form, separately can contain their own special portion of consciousness energy, the Absolute. Recall the analogy of wave patterns on the ocean. All energy could be described as consciousness (minute units) in different patterns, created by templates. We look at an advanced computer system, or robot, and know that it does not experience—the best it can do is to be programmed to respond and copy a behaviour of experience.

At the most basic level, life and the essence of consciousness could be considered as infinitely nonlinear—this is a state of true unity. Normal logic fails utterly to handle this and only deals with the external 3D quantitative level. As an analogy, one could apply logic, say, to observing two staff members of a large company as they, for instance, chat about some common interest—maybe a television programme. But this same logic won't handle the relation between, say, two staff members who are each working on a component in unison which is a part of the final product when, say, they have never communicated. Clearly the information coming down from the manager level, or higher, explains the hidden connectivity between the two (an analogy for nonlocality in physics).

Noel Huntley

We have been discussing some of the characteristics of life and the universe which tell us that the basic properties of life, experience and consciousness are not quantifiable. We use the fractal system to explain a structural system of dimensions, ordered from the lowest level (3D) of maximum objectivity in the ratio objective/subjective through higher frequencies, rates of information, with increasing integration to higher orders of organisation and finally a single source and maximum subjectivity.

The quantity 'one' makes better physics sense than some other number. We recognise (and we increase the logic by this) that there must be nothing outside this wholeness—everything inside it (there is really no such thing as 'nothing'; there are only relative 'nothings'). Thus in a sense it is even beyond 'one'. This is a very satisfactory application of tests of truth in physics, for this particular test there must be no preferences; the concept must have no, or minimal, bias, assumptions or prejudice.

This is why science needs to recognise the full application of the fractal system to the universe and creation itself. The holographic and fractal models are powerful clues for science in an understanding of how creation occurs. When we use the fractal system internally, that is, going into inner space, sometimes referred to as 'nested levels', it does have a natural termination as indicated above. It is hierarchical and pyramidal. As we move up the ranking system of a company organisation it stops at the top, the president. We can't go higher.

How does all this relate to the subject of 'something' and 'nothing'? The author's article, There is No Death: You Can't Die if You Try,[1] uses the thought experiment of imagining gradually removing all matter from the universe—all particles and waves, and then asking would there be anything left after the last particle or energy has been removed (which automatically removes space and time).

It may not be at all obvious that a resolution, or at least a conclusion of the problem defining and evaluating nothing, has crucial applications and profound implications to theories of the origin of life and the universe. The best simple analogy for what is left after removal of all matter, energy, space and time, is one given earlier. The final absolute condition of infinite potentia, all possibilities can be represented by a vast ocean. Patterns of waves and vortices manifest our matter and energy forms. Thus removal of all the ripples (structures) and whorls (particles) leaves the ocean still there. All manifestations of existence (all matter and energy patterns) disappear into nothingness. But where science thinks it is a true nothingness it is a state beyond what is quantifiable—outside the reach of science no matter how advanced it becomes in instrumentation for measurement and detection.

31.

BRINGING IN QUANTUM COMPUTING

It is necessary to create or find natural phenomena in which the coherent state of superposition is preserved, and we could have a fully working quantum computer.

Computer scientists today, involved in the research and development of the quantum computer, encounter the paradoxical difficulty that whereas in the ordinary material life the environment is predictable (when all information is known), but this information springs from the quantum realm of wave activity where nothing is known specifically—everything appears spread out. Encountering this problem is not difficult to imagine when we consider the pyramid structure that we have used regularly, Figures 10, 11, 14. At the top of the pyramid everything is merged into one—even quantum physicists some fifty years ago, as we have repeatedly stated, commented that it is as though there is only one 'big' electron.

Note the terms in use and changing preferences with some scientists and nuances of meaning, such as quantum reduction, collapse of the wave function (wave packet), decoherence, selection from multi-possibilities; they are all referring to the same process. But we shall see that quantum reduction divides into objective and subjective, and decoherence is biased towards a collapsing of the wave function that occurs apart from consciousness. Furthermore, prior to this, the quantum realm before quantum-reduction is described as in a phase-correlated state, a state of coherence, a state of superposition and entanglement, also expressed in terms of the wave function, and even the word holistic has been used. Quantum entanglement is a form of superposition. It is the underlying mechanism that enables the phase-correlations to occur.

The quantum realm was originally merely a mathematical construct but an increasing number of investigators are recognising that it has a tangible or substantial reality. We are proposing here that this quantum realm and its extension covers a range of phenomena, in particular, orders or organisation immensely complex. The wave function is, of course, also mathematical, and has been called a mathematical fiction, but as we recognise endless wave/frequency patterns underlying all of creation it becomes real.

The argument that we have presented, that the absolute is infinitely nonlinear, has a basis for the quantum reality of possible infinite superposition and infinite possibilities. In other words, the superposition character of quantum theory could have its origin in our nonlinear fractal hierarchy. Science only applies 3D logic (classical logic) that handles systems in which the whole equals the sum of the parts, and not when the whole is greater than the sum of the parts. The latter is higher-dimensional and requires at least two levels (that is, an extra higher-order layer). Thus there is the presence in creation of nonlinearity and superposition of states—with which our conscious minds have great difficulty grasping (in extreme nonlinear systems). However, we now have quantum logic, a new approach to information theories about the universe, not dependent on classical logic.

Thus science is now researching computer bit theory, but not the classical version. Quantum theory presents potentially much superior methods that are being investigated in the field of quantum information theory, which extends the binary bits to co-called qubits (quantum bits). The probability wave that describes the micro-particle in the quantum field does more than inform the particle where it could be located. It tells it how to behave in all situations; for example, when a photon will pass through glass or be reflected. Both states exist in the quantum realm simultaneously until detected but they can not only be in many places at once but do many things at once. This is the key to the quantum computer. But the problem is how to retain the simultaneousness and superposition.

Thus a newly developing interest is in the quantum computer, which appears to be a natural progression from classical computers since as the computer bit size reduces we run into quantum uncertainty—a serious problem for the classical design. Thus in the quantum state the uncertainty itself allows the superposition to exist with the potential for the development of the quantum computer. Quantum uncertainty is not a problem for conventional physics laws, since the statistical macro-level gives us the classical deterministic result. However, the quantum-state superposition is lost in the measurement or observation and we need to retain this in the quantum computer. What is 'superposition'? This is the quantum-realm condition prior to the collapse of the wave packet (caused by measurement, observations and many natural interactions).

The requirements of a macro-quantum computer are to delay the observation consequence that collapses the quantum-wave function, otherwise ordinary observations would reduce the quantum computer to the classical computer in selecting a single probability before its superposition function could be used. This is the problem. Thus if we could exploit the quantum phenomenon before the collapse, of which the latter would select the single possibility and destroy superposition, we could have a macro-quantum computer.

Quantum physics has long since revealed that objects can be in two states at the same time—in fact any number of states. The well-known two-slit experiment reveals the wave nature of light by producing interference fringes. But much more astonishingly and highly significant, it reveals that single particles, that is, one at a time, for example, photons, can create the fringes, forcing one to consider that a particle has gone through both slits simultaneously (called superposition) to interfere with itself and cause fringes to appear (rather than concentrated dots on the target screen as expected from single beams).

This has developed into the conviction in quantum theory that all objects in the universe are capable of being in all possible states. However, we cannot detect them in this state. These states will be reduced to a subset of the states when observations occur, which includes measurements and, in general, certain environmental interactions. This is the collapse of the wave function giving the single possibility that we observe. However, there are other kinds of interactions that preserve the quantumness (the quantum phenomenon of being in many states at once, referred to as superposition).

Decoherence enables us to receive the solution from the quantum computer but the computer's simultaneous calculation will not be revealed—it can only give one final result. Therefore it is limited to single-answer type problems. When the quantum computer is given a task it divides it into many versions of itself, each works on a separate piece of the problem. Then the pieces come together and a single answer comes up. Thus currently it would seem that the quantum computer is limited to handling a very few problems.

At quantum sizes it is easier to isolate chosen small systems that are less likely to collapse the wave to our classical, conventional result. In the larger scale there are many particles and interactions that will reduce the wave character to the ordinary everyday object level. Common everyday interactions may destroy quantumness to give a definite answer, a single possibility. These interactions within the environment act like a measurement, bringing the collapse of the wave function (from the quantum realm of phase-correlation, a coherent state, to phase-randomisation, a state of decoherence). When we make measurements or record the location of, say, a particle, it is always in one position or another; never both. However, if we do not measure, but instead interact with the object in a way that does not record its location, then the object behaves as though it is in both places at the same time.

It is an important area of research in quantum information to find phenomena in which the coherent state of superposition is preserved—seemingly in the quantum realm. Initially this has been revealed at very low temperatures (the familiar superconducting conditions), but more recently examples for possible application have been revealed at room temperature.

A striking example of a macro-system exhibiting superposition and coherence is the well-known Bose-Einstein Condensate.[2] Originally, physicist Bose devised rules for describing when two photons were the same or different. Einstein extended this to atoms. In the Bose-Einstein Condensate thousands of atoms of a gas at very low temperatures will coalesce into a single 'superatom'. Atoms of a few elements cooled by laser bombardment (slowing down the fast atoms) will drop to their lowest energy state and act like a single giant matter wave, and look a little like a drop of water. This can be achieved now with a few million atoms, and presents us with a new behaviour of macroscopic quantum matter.[3]

The atoms of liquid helium at very low temperatures will all go to the same lowest energy state and then all act like one big atom. The helium atom has boson characteristics (can interfere and merge) at these temperatures, that is, can crowd with other bosons and not be kept apart by the requirement of separate quantum numbers as is the case with fermions, such as electrons, most atoms, protons and neutrons. Such experiments can reveal alternate collapses and recoveries of the coherent matter wave.

An entangled state of many-atoms, such as this, could be used for quantum computing. Thus quantum superposition phenomena can occur on a macroscopic scale. For example, quantum superposition has even been demonstrated in a piece of salt at low temperature, and quantum coherent states have been found in microtubules of neurons in living organisms.[4] There are further applications utilising ions and photons, or the spin characteristics of certain particles.[5]

We have an interesting comparison, or it might be called an analogy, for this requirement of delaying the decoherence in the earlier discussion. It is the phenomenon of the second-order quantum reduction in Section 35. We mentioned the resonant condition (coherence) of a falling body prior to its being arrested in striking the ground (or interfered with by any physical action), that is, between the first and second-order quantum reduction and the need to utilise this condition and not the state after the second-order quantum reduction—due to action, giving loss of the group (atoms) coherence. The falling body is part of our reality (first-order reduction or collapse of the wave has taken place giving the first stage of decoherence) but the body is still in a coherent state as a group with space until physical action on it causes the second-order reduction of the body, subsequently destroying its (group) harmonic one-bodied relationship with space-time (recall the one-bodied craft-propulsion system, Figure 5). This means that just as quantum theory tells us that only when an actual measurement is carried out on a particle can it be said to possess momentum, our second-order coherence presents us with examples in the universe and our immediate environment in which macro-bodies do not possess momentum, inertia or kinetic energy until a measurement (physical) action on them is made. We shall come back to this as we focus on the learning pattern as our prime example in illustrating mind-computer coherent quantum states of superposition and superimposition.

To recapitulate, current conventional (classical) computers use the binary system, utilising states '0' and '1' and the basic element, the transistor, provides two voltage states for these two bits. These represent the 'on' and 'off' condition in an electrical circuit, respectively. The bit can only be in state '1' or '0' in any given moment. In quantum information, however, it can be in both states simultaneously, and is called a qubit. The more qubits, the more superposition of simultaneous possibilities there are. However, this is in the quantum realm of coherence of quanta, in which when we measure a qubit, we reduce it to the classical result (one outcome). In fact with qubits we can have not only a '0'or '1' at the same time but an infinite number of states between 0 and 1 by using probabilities.

Since computer transistors are utilised in silicon chips, and semi-conductors (the medium of silicon chips) function entirely quantum mechanically (and can't be understood classically), it would seem such materials are the natural next step for researching the potential of qubits. A quantum computer could check multiple possibilities simultaneously and exploit the principle of quantum superposition, and compute the result, a single outcome. However, as we have indicated, in larger systems (macro as opposed to micro, the quantum realm) the probability of the system to decohere as a result of interactions with life and the environment is very great. In effect, information leaks out. The more atoms there are in a superposition the more likely it will decohere to the environment. And the classical situation takes over. Thus the larger the system the more ways there are for quantum information to leak out and superposition is lost. Interestingly the uncertainty of the quantum level is a requirement to maintain the superposition, and when we try to create certainty we lose the superposition.

In terms of quantum information theory, it seems one must have uncertainty otherwise we would alter the behaviour and properties by interacting, by measuring, by decohering the quantum state—we would destroy the atom's or particle's wave behaviour, such as the interference pattern and there would be no quantum theory. Thus the uncertainty principle appears to protect the quantum world, which in turn protects our world. Not so surprisingly, since as we look closer into the microscopic world it gets fuzzier—nature it seems does not allow us to be too accurate or accurate at all in the uncertainty principle applications. Look at our Figure 12 again showing waves within (or carried by) waves. What it means though is that the lower-frequency packet at the bottom only detects its counterpart in the higher levels, which will appear blurred out, less distinct, if we imagine a microscope examination.

Quantum information tells us that at the particle level, the basic level for existence is inherently random, but can in some cases also be deterministic at the macro level—not necessarily statistically—and free will is being considered to be somewhere between randomness and determinism. That is, quantum information tends to indicate that determinism can arise from random origins. However, our reaction is that this is so much on the surface of reality and observation that such a conclusion can't be justified. There is so much hidden activity. We see in the later section that Figure 15 indicates randomness itself will be relative. A higher order impinging into the lower fractal levels manifests as complexity, and either the order cannot be intelligently evaluated or it appears to us as natural randomness and disorder at this lower level.

True quantum randomness is considered to be an excellent mechanism for preventing the universe from dictating events and overruling life. Thus quantum physics uncertainty could be an essential factor in the proper functioning of a universe containing life. It isn't that the universe is protecting itself but that it is providing a multidimensional structure of possibilities, graded for man to select by interaction. Later we discuss free will and how this relates to physics developments.

The Emerging New Science

Nevertheless, it doesn't matter how clever we become mechanistically, or using whatever subtle devices we have (such as from quantum theory or quantum computing to explain life and existence, behaviour and apparent free will), even these quantum-theory approaches will never handle the experiential sensation aspect—that feeling of 'aliveness'. However, a good point in favour of the New Science is that academic quantum information interpretations agree that reality is context-dependent—one of the principal features of life in a universe. We further add that all energy, knowledge, including the results of scientific methodology, are contextual—Section 21.

Although quantum information is a valuable tool for providing an underlying description of reality, and clever though it is, it rests totally on what has been called tried and tested physics. As we have seen, science is based on limited closed contexts, but quantum theory is capable of bridging our closed-minded knowledge with an ever-expanding context into unimaginable sciences. Quantum theory has greater potential than to just be utilised at the 3D quantitative level.

Thus quantum information as it stands is still simply another interpretation of existing limited, relative physics laws. These interpretations not only ignore the absolute qualities of existence (leaves them out) but still fails to recognise the underlying depth of reality, the infinite eternal nonlinear system indicated by the fractal, holographic hierarchy (Figure 1). In fact we may come to understand that, for instance, the holographic configuration gives us a clue to the interface between absolute and relative. Structures such as holographic and fractal are natural by-products, and are 'attempts' by energy, space, and time, to mimic the infinitely nonlinear, absolute potentia characteristics.

Quantum information indicates that having fundamental randomness would appear to limit theory and laws. But as we shall see, the potential relative aspects of randomness, as depicted in Figure 15, Section 33, is not an immediate problem, or one at all.

Quantum theory suggests that qubits come from nowhere and there is no prior information required in order for information to exist. This is not particularly surprising since this type of information is insubstantial (non-physical). This feature is clearly illustrated for the reader later when we use quantum information's teleportation example. It is interesting how science will give a reality to quantum information but not the non-physical (or energetic) origin of existence.

As per quantum information, this use of the qubit apparently breaks the infinite regression chain, that is, to always need a more fundamental law to explain the current one (the problem of getting outside the context). However, the whole argument is limited by analysis itself, which is quantitative and quantifiable. This academic logic automatically ensures that the investigator will never be able to see the forest for the trees (while inside the forest). Science is desperately attempting to explain great complexities within a simple framework and is constantly unknowingly quantum reducing in a second-order manner its own observations. We shall take this up later.

Having now gathered a certain amount of information on quantum theory application, let us give a brief summary and history here of basic progress in this field. As we indicated earlier, the basis of quantum physics is the concept of indeterminism, that is, uncertainty; whereas in classical physics if we knew the initial properties of particles in the universe we could determine the outcome of all future events. Thus classical physics tells us that the universe is deterministic, and quantum physics that it is inherently uncertain, with a very high degree of uncertainty at the micro-level. However, at the statistical macro-level, the uncertainty is reduced to being negligible and we can have determinism again.

The chaos theory outcome is different again. Uncertainty arises again now at the macro-level. A classical system that is sufficiently disturbed, such as flowing water reaching turbulence, the chaos condition, causes uncertainty, even with apparently complete data. The tiniest change at even the micro-level can amplify up to macro-proportions. But here we are interested in quantum-information contributions.

Now, experiments with photons and mirrors seem to indicate that determinism can arise from random origins, to which we alluded earlier. Again, from our point of view, there is a neglect of information going on in the background, outside the

The Emerging New Science

resolution of scientific instruments. The purely quantitative approach of science and quantum information, according to the latter, is telling us that we are all just information. This is highlighted with the teleportation thought test—mentioned later. The bizarre features of quantum theory are being accepted more today, even proven in some cases. Prior to the observation or measurement, the quantum realm is in a holistic, phase-correlated state of superposition—all possibilities of any given outcome exist simultaneously (hence the famous 'Schrodinger's Cat' analogy). The collapses of the wave function (selection of a single probability by observation) is a crucial example, but is not inherent in the equations. The collapse idea is a deduction from simple observation, everyday experience. It is due to some mysterious process in the interaction between the material or physical world, and the quantum state (that is, between our linear existence and the quantum realm of superposition of all possibilities), and does not yet have a complete answer, as science well knows. But this so-called 'collapse' is still a good explanation, even though we need an understanding of the mechanism. This is the Copenhagen Interpretation proposed many years ago, which was met with scepticism by many scientists. Nevertheless, it has stood the test of time and as an overall concept, already exists (not just as a result of quantum physics) conceptually and independently within less scientific sources—less conventional areas of more experiential knowledge, such as New Age, spiritual sciences.

The key cause of these paradoxes can come from the source of the fractal holographic structures of creation; that source being the infinitely nonlinear Absolute, where everything is everywhere at once (from our linear quantitative view). Amazingly quantum theory continually reveals this, but science desperately endeavours to explain everything on a one-level basis—such as with the comparison of our computers (or scientific knowledge generally) with the ground-floor workers of our company organisation analogy, compared with the mind computer containing the 'higher ranks' as well.

We have seen that the quantum realm, rather than having only mathematical significance, can have a substantial basis as the infinite absolute of no motion, no particle, wavelength, space or time. It has no location and is eternal. But to increase the reality of science's quantum realm we need to fractalise it into different orders, dimensions, each level having its own degree of randomness. Its dimensional framework utilises both deterministic events and randomness—an already existing conclusion of quantum information regarding the requirements of free will. Simultaneity and superposition are conceptually not a problem with the fractal, holographic model (mainstream science avoids the nature of unity or holistic systems, which is a contradiction).

Thus modern science, expanded by quantum computing, gives us the binary system (0, 1) for processing information (a one-level system, where classical physics necessitates the either/or condition of the bits 0, 1), whereas quantum information permits the simultaneity of these states, as already described. However, our new model gives a much expanded application of the computer bits, extending the concept into the fractal system and handling unity—which is what a true quantum state has anyway (unity/coherence).

The new model utilises the triad principle. We can still consider bits (0, 1). Let's designate them A, B, except that we don't need the '0' (this is the electrical 'off' state, zero current), since life and the universe can be found to operate on geometrical intelligence as already mentioned—it is all hardware in the form of energy and frequency patterns. In computer software, '0' is a 'thing'; it represents a physical state (absence of a charge) and is equally effective as '1'. Thus the computer bits 0, 1, have equal informational value. Let us use A and B as the equivalent of two bits. Similarly our A and B have equal value (status). Recall Figure 13 showing emergent software and quantum regeneration. We need to incorporate this principle.[6]

Figure 14 shows the addition of the third unit C. This unit integrates A and B and regulates the relationship between them. It can instruct A and B to be in sequence at a particular rate or it can instruct both A and B to act simultaneously. These are elementary programmes. One must envisage countless numbers connected pyramidally. We see the same principle in Figure 11 on a large scale. The three basic units representing the smallest

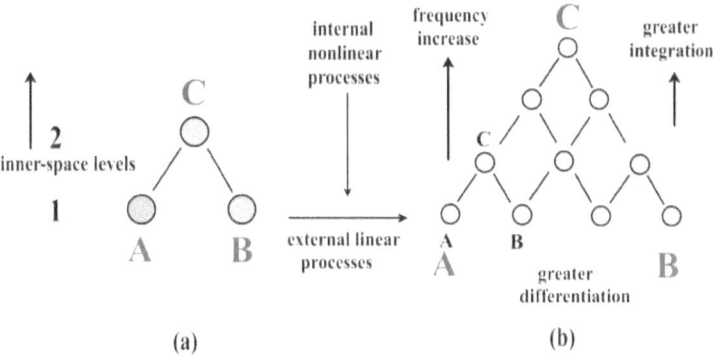

FIGURE 14: Triad principle of nature's creations

meaningful element of computer information as a whole could be considered to be the fundamental mind-computer element—a triad. Element C represents both quantum regeneration from A and B and the prior condition in the formation of A and B before fission from C.

An example of this principle is how learning patterns are structured that provide all physical muscular movements in living organisms.[7] We can see that the fractal system is inherent within it—also it is intrinsically holographic. [This was the chosen study for the author's experimental psychology doctorate dissertation: The Mechanics of Learning: A Holographic Nonlinear Theory.]

We might note that the extraordinary features of quantum theory or quantum information, such as nonlocality, superposition, simultaneity, are inherent within the fractal, holographic mind-computer system. This sounds encouraging for quantum information but the latter is being restricted, as with all science, to the quantitative, linear level, which in Figure 14 would correspond only to A's and B's and no C's. Mainstream science does not even postulate upper levels (nested internally in inner space), and subsequently the framework is much too narrow.

The term 'superimposed' better describes these nonlinear inner-space layers since they are placed one on top of another (or at least overlapping). Remember 'superposed' means super-positioned, for example the unobserved particle is everywhere. This relationship can be called 'horizontal' as opposed to 'superimposed' in which layers can be considered placed 'vertically'. Science only recognises superposition, the flat, linear, one-level of order.

We have mentioned quantum regeneration (not recognised), the opposite of quantum reduction, a very important process. For example, the laser design creates the laser beam of coherent light; the key word, 'coherence', in quantum information. But scientists don't recognise they have created a quantum state of coherent light—a single 'collective' of the original independent rays of light, quantum regenerated (this is where the power comes from). This single 'collective' is a new state, a coherent entity from the quantum-realm virtual state. What we are implying here is that quantum computing may be able to utilise the principle of quantum regeneration to produce stable coherence and manageable qubits. The laser is an excellent example (and possible application). If the coherent state is collapsed, it will instantly be regenerated by the laser device, exhibiting alternate collapses and recoveries, as required.

Note that we quantum regenerate a coherent quantum state when in a state of art/music appreciation, but 'destroy' the state when analysing (collapse of the complex wave); in relationships, for example, two people can be on the 'same wavelength' (agreement, coherence, a collective state), but destroy it by argument; in coordination where we create stable learning (wave) patterns, these are quantum reduced to lower levels in converting nonlinear information into linear information (such as finger sequences on a keyboard)—see later sections; finally, the advanced spacecraft achieves a stable coherent oscillation by quantum regenerating (entraining) all the atomic oscillations of the craft, and simply turns it off when not required (Figure 5).

As a final comment and recap we have the development of artificial computers that follow classical and linear characteristics; they compute on one problem at a time but very rapidly (one can of course have parallel sets of separate solutions going on simultaneously). Concomitant with this, science teaches and emphasises the brain model for the intellect and consciousness. Note that our artificial electronic computers are far faster than the brain but not the mind. Nevertheless there has been early research in the development of quantum states for faster computers—these are instantaneous states. Recall earlier reference to Professor Terry Clark's success in this important field. Today we have the first glimmerings of the quantum computer, that can compute on more than one instruction at a time, that is, utilise the simultaneity and superposition of quantum processes.

The irony is that man is not limited to the brain and in fact already possesses an amazing mind with quantum-computing capabilities—could a physiological brain research and develop a quantum computer? Emphatically, No! We shall take this up later with particular emphasis on the application of the learning pattern to this topic.

Notes:
1. www.nhbeyondduality.org.uk. Article: *There is No Death: You Can't Die if You Try*.

2. http://physicsworld.com —*Bose-Einstein Condensate*.

3. Vlatco Vedral, *Decoding Reality*.

4. Ibid.

5. M. Chown. *Quantum Theory Cannot Hurt You*.

6. N. Huntley. www.nhbeyondduality.org.uk. Article: *New Education Series, The Triad Principle of Creation*.

7. www.nhbeyondduality.org.uk. Article: *The Learning Pattern*.

PART SIX

QUANTUM REALM

32.
INTRODUCTION

Is quantum theory complete, or is it a subset of a much greater paradigm?

The quantum realm can be considered to be a state underlying the material world, but quantum mechanics is basically mathematical and our quantum realm inevitably has a mathematical framework. However, quantum theory conveys more realism than is recognised in official science. A particularly striking feature is that it is a framework with infinite degrees of freedom for the quantum field. Thus the quantum field forms the basis for elementary particle physics and field systems, and underlies all creation. The quantum realm has wave properties, but quantum waves will only be detected as particles in the material world or indirectly as waves, such as in interference experiments. We use the term 'quanta' to describe energy in packets or bundles, observable in our material world, such as in light or, in general, radiation. When measured, these are discrete units of energy and frequency.

In quantum theory, particles such as electrons, protons, are recognised as having both wave and particle properties. The particle (or many particles) can be represented as a wave shape called a wave function. In the quantum realm, matter and particles are considered to exist as described by waves of possibility. Quantum theory at the particle level can't predict the precise particle event of what a measurement gives, or observation, only the probability. As soon as we try to measure the quantum wave it collapses from many possibilities to a single determination, the particle mode, and the word 'quanta' is used to describe the discrete energy states—not continuous as at the classical level.

From our point of view, quantum theory is telling us much more than is apparent and provides the needed link into higher strata of knowledge and scientific discoveries—it indicates a much greater order of complexity. For example, nonlocality is not some magical incomprehensible process (a common scientific view, though not admitted). It is a natural result of a much more complex and larger 'quantum realm', which goes beyond the quantum vacuum, in other words is a bigger picture.

The quantum world within atoms and particles, when measured or observed, reveals discontinuous energy, whether energy levels in atoms, which are standing waves, or discrete units called quanta, as appears in radiation. If an electron is freed from, say, an atom and thus free to move in space, in the quantum realm, as given by Schrodinger's equation, the wave will spread rapidly (like water waves do) and overlap and interconnect, becoming coherent with, other waves. Thus its location is uncertain but we can find the probability of it being detected at a particular place (however, one shouldn't assume the particle has a location in the quantum realm). Regarding larger masses, the quantum wave spreads extremely slowly and the object's location or properties of the particle can be determined accurately. Atoms and particles can be described by waves, thus superposition is possible in the quantum state, enabling the atoms and particles to be in any number of places at once. There is no limit to the number of waves that can be superposed but there is an inherent uncertainty until observer interaction occurs and the selection of a single event is made.

Let's say an electron was freed from the atom by means of a collision in the quantum realm. Strictly, the collision will result in the electron's quantum wave being both free and still connected to the atom, called quantum superposition. Both states exist simultaneously. We don't detect this, and the measurement or observation manifests only one or the other.

Science's quantum mathematical realm could be re-pictured by us as a border level between the 3D and 4D spectra of the astral plane but, in fact, quantum theory penetrates deeper with its concepts of infinite possibilities; with its 'holistic' framework, and properties, such as nonlocality. The collapse of the wave function from infinite possibilities will be limited by

The Emerging New Science

free will restrictions built/programmed into the wave patterns both in life and the environment, causing selection. Quantum theory's infinite possibilities would take one to the absolute of complete free will with no restrictions (see section on free will).

One may encounter the term 'quantum vacuum', representing the fundamental state of the quantum realm. It is referring to the quantum state with the lowest possible energy (generally described with no real particles present). Thus it is not empty but contains the energetic activity of electromagnetic waves and virtual particles that appear from it fleetingly and disappear. The virtual particle is non-observable, compared with, say, observable electrons, protons. The vacuum is theoretically achieved when one removes all matter and lowers the temperature to absolute zero. This is the classical view giving an apparent true zero, hence the misnomer zero-point energy. However, as indicated, it was discovered that the 'vacuum' was thriving with energy.

Virtual particles also provide the forces between observable particles. The virtual particles, such as photons, are exchanged between real particles, for example, electrons. Thus the force effect between observable particles is accomplished by exchanging virtual particles. The 'boson' virtual particle governs the weak electromagnetic force and the 'gluon' virtual particle governs the strong force (nuclear). There is no established particle yet for the gravitational force—though the 'graviton' virtual particle is a candidate. The fourth force is a weak nuclear force involved in decay. These interactions are perpetual in this quantum vacuum. Virtual particles include their antiparticles, which are opposite in sign. The pairs continuously annihilate each other. It is happening all the time. The particles and their anti-counterparts have been referred to as mini-black and mini-white holes in the quantum vacuum, hence the expression quantum foam to describe its appearance.

The quantum foam is conceptualised as forming the foundation of universal space. The perpetual virtual particles appear and disappear as they annihilate one another, with measurable energy—recall the ocean analogy. The uncertainty principle allows them to 'magically' appear, and during small scales of space time to disappear. Even large fluctuations of energy are allowed during these short intervals without violating the conservation of energy. The smaller the scale of space and time, the higher the energies of the virtual particles, giving rise to small scale curvatures in space time—hence the theorised appearance of the foam-like nature. These vacuum fluctuations are accepted as nothing more than modulations of space, giving the vacuum energy or the zero-point energy. These micro curvatures of space break through the 'surface' and create the mini-black and mini-white wormholes.

Thus we shall use the term quantum realm as it relates to quantum mechanics but concurrently in the New Science we shall be regarding the official quantum realm as a subset of a greater multi-field and space-time configuration of different orders. In simple terms, this is the vast ocean; the analogy we have used to express the infinite, absolute, nonquantifiable continuum of all possibilities. Thus nonquantitative states such as: the experiential, basic consciousness, or sentience and awareness, the 'aliveness' characteristic, all are considered in the New Science to spring from this 'ocean' potentia. There is nothing outside it and its most natural and intrinsic inherent expression appears to be in geometry and frequency patterns. Schrodinger, one of the founders of quantum theory stated: What we observe as material bodies and forces are nothing but shapes and variations in the structure of space. Compare our extreme example of this in the ocean analogy. However, between the nonquantifiable ocean of 'nothingness' (the most basic state) and the official quantum realm, is an extensive fractal, holographic, higher-dimensional structure of increasing degrees of order—like inner-space nested levels.

The fields of the quantum realm are composed of quanta (called field quanta—possessing discrete quantities of energy, not a continuous range), which are like ripples in the field and simply look like particles when observed—they are all basically indistinguishable; recall that the virtual particle is a theorised non-observable particle expressed in quanta. Matter appears as made up of separate particles but in terms of quantum physics underlying this, the quanta are in a dimension in which nothing is separate. The quantum realm is coherent and holistic—there is a nonlocal connection between all quanta. Everything is interconnected in the quantum realm and any one part cannot be described without considering the rest. The property that fractals exhibits is known to be present at quantum scales but there are macro dimensions within, though beyond the limited scope of current quantum physics.

In the New Science, quantum vacuum is basically a scalar field, an electrostatic potential, that is, a state without mass: real particles, photons, electrons. The force property of the field only arises when mass is present, and even then such particles need to be constrained or artificially moved externally in order to 'express' their Newtonian characteristics: mass, force, inertia, kinetic energy. This is not recognised by mainstream scientific thought. A scalar field is not a force-field, as is the electric field, unless mass or real particles are present, and even then such particles need to be constrained or acted upon artificially, or by random forces, and moved to invoke the force (but there is no force in 'free fall'). The scalar field is a coherent field state (as is indicated by quantum-field theory itself). Similarly, as we mentioned earlier, a body in the gravitational field in free fall is not subject to the gravitational force—it is coherent with the flowing space-time gravity pressure. The gravitational field is not acting on the mass with a (separate, external) force—which would be a decoherent condition—until the free-falling body's motion is arrested externally, or when pushing it, resisting it, or letting it rest on the ground. Note that Einstein recognised a falling body was weightless, in particular, when we experience personally jumping from a height. Unfortunately he allowed the unbending authority of Newton's laws to dominate (force merely cancelled by inertia at the acceleration due to gravity) and missed the clue that the body was coherent with space time.

As a final comment let us emphasise again the potential similarity of this model to the ocean analogy in which all of creation is nothing more than patterns and curvatures in the infinite quantum realm. In the view of the author, this abrupt change from the quantum realm to the material world is a fiction. There is not just the primary collapse of the wave packet, from the supposed coherent state to decoherence, but secondary quantum reductions, as we shall explain in the proceeding sections (Figure 26, Appendix D).

33.

QUANTUM REDUCTION AND QUANTUM REGENERATION

There is a continuous series of quantum regenerations and reductions that quantum leap up and down the energy levels of the fractal frequency spectrum (Appendix D, Figure 26).

We can initially think of the quantum realm of all possibilities as a background of wave forms to our material existence, and all waves are radiating out to the universe. Consciousness interacts with this quantum realm with subjective reduction (decoherence) to a single probability. Note that it is called objective reduction if one believes the collapse causes consciousness. Figure 12 takes the view that it is subjecttive but consciousness and interactions are continuous and everywhere (the latter tends to favour the objective view). There is no fundamental objectivity; a specific form of the universe is not already there without life and consciousness, but everything contains this consciousness and the interactions are making the selections. The basis for the resolution of this question of an objective universe is covered later under the Einstein-Bohr debate. Remember, interactions and observations, are contextual, meaning that when anything is measured or observed, including many more natural interactions, the resulting universe view is in the context of these sources.

There must be a natural, or common agreed upon, quantum reduction to provide the lower extensions of the dimensional hierarchy. The fractal boundaries are fixed and predetermined. For example, our 3D might be the lowest universe or space-time fractal. But we have seen that there is also quantum regeneration, the opposite of quantum reduction. There is freedom between the fractals; for example, a simple analogy might be that the various parts of our arm (a fractal

The Emerging New Science

system of joints) can be moved relative to the fixed joints. Between the fractal levels of existence, by putting order into the parts (elements of existence) one integrates greater wholes (larger quantum-wave packets). Thus life within a particular level can select the quantum possibilities in a manner to expand consciousness by quantum regeneration—extending the total wave packet (until it qualifies for the next fractal level).

Figure 15 can be a useful guide to bring together many diverse features: principles of the universe, in particular, interpretations in science, including modern ideas on quantum information processing. The reader may see a resemblance to the idea of using letter-ordering with the intelligence test in which one has to spot the order of repeating letter groups.

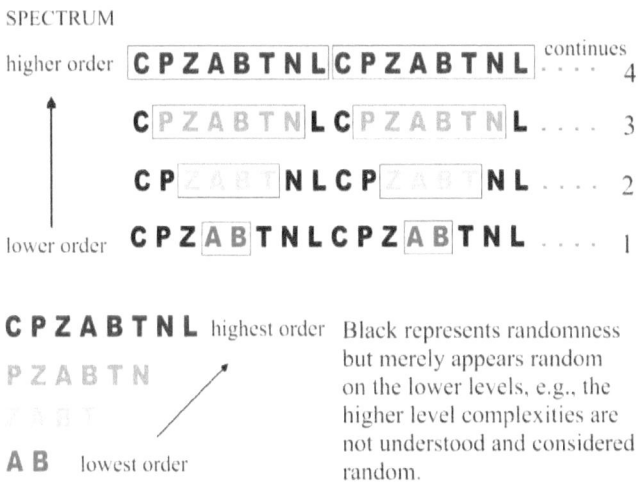

FIGURE 15

The figure gives the reader some idea of higher order, complexity and randomness. The higher order at the top is a complex modulated wave and is in a whole quantum state—that is, all waves are unified and in harmonic resonance (not all the same frequency but mathematical/harmonically related). We are referring to a modulated wave as a wave function (originally in quantum theory this was purely a mathematical description).

A particularly interesting feature of Figure 15 is that it demonstrates complexity theory of mathematics in that it illustrates the idea that the higher the order, the greater the complexity of the elements and the more random it may appear—at the lower level. It may be more difficult to spot the order. This applies in the extreme in our 3D existence. How many people can appreciate the high order in great art, or recognise that a scientific law may be of a low order and needs to be updated and upgraded.

Contemplating current scientific procedures for acquiring knowledge we might think that one of the goals of science, when quantifying observations/data, was to seek out quantitative methods that would, as a primary action—not just an apparent consequence—eliminate all subjectivity, such as consciousness, life, the experiential. This unfortunately is now happening in quantum physics, which makes it more acceptable to the materialist.

Now the real complication arises when we recognise that between our material existence—detectable by science or physical senses—and the eternal and nonlocal potentia in its primordial state, is an immense complexity of ordered and random structures based on fractal and holographic designs. Levels 1 to 4 in Figure 15 can represent fractal levels, spectra or wavebands, or increasing dimensions of an existence, or 1 to 4 could be within 3D, representing different degrees of complexity and order. For example, a scientific law or principle might be based on the order AB—a detected consistency that enables a formulation to make predictions, and supply information about some phenomenon. It may subsequently be recognised that this scientific theory does have some weaknesses, fuzzy edges or anomalies (these may, however, be blocked off by the scientists through wishful thinking or arrogance). These weaknesses may be due to missing factors represented by Z and T, P and N, C and L.

One might spot that law AB needs to include T but that doing this causes loss of symmetry that contains, say, AB in its limited context. As a result, T tends to be ignored since symmetry is a powerful 'tool' for identifying truth. In this case, there is a failure to recognise that the loss of symmetry in adding T is simply a temporary condition leading to a greater symmetry as Z is discovered. Anomalies can be powerful indicators of higher orders.

Thus as we evolve our existence, and concepts of a higher order become available, say, level-2, we find the need to update our laws. In this analogy, this means at level-2 the random aspect of ZT is now recognised as basically due to a higher order, and is now incorporated in the higher-order formulation ZABT.

At the highest structure of creation, say, the top of the letter-orders in Figure 15, the basic guidance/command might be to simply exist, and within this the drive to survive, not just for the individual but for the whole. We might note that at the most basic level a drive to be 'good' would arise from any emphasis on survival for the whole.

The group AB represents what would more easily be detected or evaluated. Bringing in, say, the adjacent Z and T could simply be extending a theory by making new discoveries or refining something such as a formulation. We may see that the diagram indicates different orders of randomness. The letters Z and T would be a lower order of randomness, and C and L the highest. This randomness can merely refer to regions unevaluated by science. It might also refer to randomness at the particle, micro-level, which we now know is quantum randomness.

Remember that quantum randomness can't be understood or predicted no matter how much information one may have about the system. It is inherently uncertain. However, when we look at large-scale events (everyday experiences of people) a statistical evaluation of the uncertain quantum randomness of particles reduces to the classical measurement or experience, giving certainty. At the quantum level, a single electron may or may not get through a barrier in a silicon chip (it only has probability), but at the macro, classical level a tennis ball predictably will continue to bounce off a wall rather than occasionally go through it. Nevertheless quantum probability indicates that, although very remotely, there is a slight chance of the tennis ball penetrating the wall.

Thus the statistical result (taking an average of many) gives a deterministic outcome of certainty. We can speculate that some of the random patterns may carry higher-order statistical outcomes, which are known at the higher level.

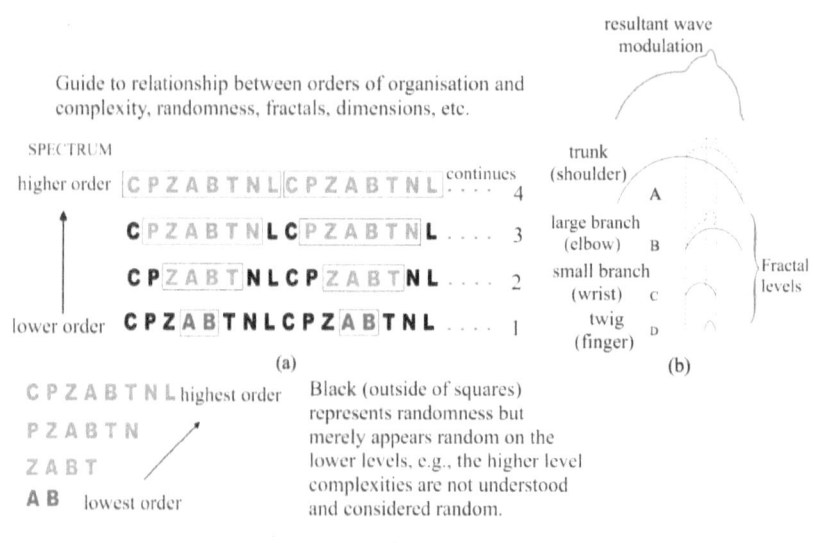

FIGURE 16

We can now introduce the earlier Figure 12, which we have placed in Figure 16, to relate it to Figure 15 and expand our concept from a principle to include the hardware, that is, the wave forms, and examine a mechanism to explain the precise process of quantum reduction from the coherent quantum realm to our single decoherent selection of everyday existence.

We are proposing that the process of decoherence (collapse of the wave function) is not only a continuous activity forming the basis of our selected reality (primary reduction recognised by science) but that this process is active continuously through the different dimensional levels, the fractal strata shown in Figure 16.

We have seen that in some of the acceptable theoretical conclusions of quantum theory, decoherence can be caused by not only the experimental measurement but all kinds of interactions. The original realisation was merely related to experiments and was termed the Copenhagen Interpretation. Then all observers became in question regarding sources for collapsing the wave packet in everyday living. This question rapidly expanded to, 'What about a mouse?' Then further, 'An insect?' All life it seems is continually bringing about selection/-decoherence from the phase-correlated quantum realm.

34.

RESOLUTION OF THE GREAT EINSTEIN-BOHR DEBATE

Einstein's biggest mistake.

In the early days of the advent of quantum theory some 80 years ago the greatest physicists debated endlessly how quantum mechanics should be interpreted or whether in fact it was complete. There were problems with the measurement uncertainty, also where the quantum world ends and the classical begins, and the plausibility of nonlocality and superposition.

The greatest debate, as recognised today, was between Albert Einstein and Neils Bohr,[1] which continued relentlessly for some 40 years and was never concluded. It still continues today amongst the more recent generations of physicists. Bohr accepted quantum theory's Copenhagen Interpretation, but Einstein, although acknowledging the unrivalled success of quantum theory, could never accept the idea that prior to the observation or detection process there was no objective world, that is, one set apart from the observers. However, the resolution lies in a much more expansive and higher-dimensional paradigm, outlined in this book. On the basis of the fractal holographic universe and its relationship with consciousness, as described earlier in discussing the original creation, we could tentatively state that both Einstein and Bohr would be essentially correct.

How can they both be correct? Recalling the description given earlier and associated with Figure 11, by imposing degrees of 'not-know' a hierarchy is created extending ('down') into

The Emerging New Science

increasing degrees of objectivity, separateness and unconsciousness. However, we know that at the higher level, the origin, both sides, observer and observed, are known and are from the same source. Consequently there is no true objectivity, only objectivity at a limited level in which the environment is unconsciously created (at the higher levels on a fractal gradient). Thus at this low 3D level there is an objective universe out there formed of different contributions at the different fractal levels by individuals and group-agreement effects—in support of Einstein.

At the same time taking into account the larger picture with no unconsciousness, there is no objectivity, all is totally subjective—and Bohr is correct. Objectivity is relative. What may be objective to us could be subjective (or less objective) to a higher-order world viewpoint. Thus the answer lies in the subjective/objective relationship. Science does its utmost to eliminate subjectivity—it is simply not understood.

Thus Einstein and Bohr were both correct relative to their own context (but they didn't know what that context was, otherwise they could have invented a theory to combine the two contexts into a higher one of probably greater truth (and we have the triad principle again). Einstein's context was emphasising and assuming objectivity as absolute (that all separation is real); that observer and observed are inherently separate and there is no other higher context.

Bohr's context was more subjective and allowed the observer to be 'outside' (or in a higher status to) the recognisable universe, thus unknowingly bringing in the hierarchy to the level of the absolute in which both sides, the observer and observed, come together as one. Thus strictly, Bohr was more correct than Einstein, since ultimately all objectivity (separateness as a reality) is an illusion. Nevertheless Einstein was very much correct that quantum mechanics was incomplete.

If the reader prefers a simple analogy which visually shows the resolution of the debate, whether the universe is truly objective and whether quantum physics is complete, we can take a strip of paper and bend it as shown in Figure 17. It is clearly one whole piece, no parts separate. Keep in mind that if we cut it up and, say, slightly separate the pieces, this would represent a degree of objectivity introduced (because of the spaces). If it is one whole, as it is, this would represent 100% subjectivity (and vice versa if we cut it into an infinite number of tiny pieces it would be in a state of maximum objectivity)—or 100% objectivity for infinitesimally small mathematical points.

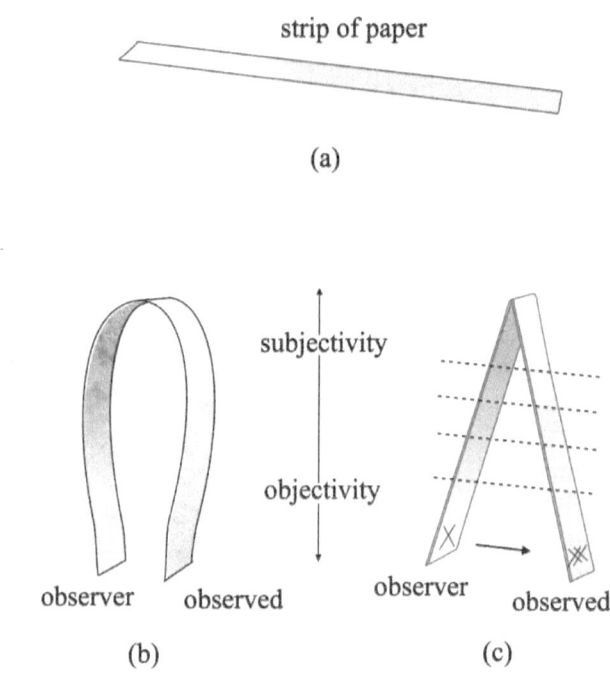

FIGURE 17: The Einstein-Bohr great scientific debate resolution.

The paper is bent so that there is a space between the two ends at the bottom. We can put a mark at A and B on each of these ends, facing one another. In the big picture nothing has really changed but if we put in restrictions as shown by the dotted lines and imagine the lower part is isolated, then we have a degree of objectivity. The ends of the strip of paper are apparently separate. One side could be a viewpoint (a person) and the other a receipt point (an element of the environment/ universe). Without the barrier, B is not separate from A. We can see that this is so by going to the top where the bend is in the paper.

Thus we can state quite definitely that there is an objective universe out there but only relative to the degree of objectivity or unconsciousness. To the degree that the observers have an unconscious mind, an objective universe is apparent. If we move up to the next dotted line in Figure 17 so that the universe level is between the first and second dotted line (this could be the next dimension or universe level above our 3D, called the soul level—just above the astral). The ratio of subjectivity to objectivity would increase. This would mean the individual observer would now recognise that he or she is aware of participating more in the environment mind-wise.

The Copenhagen Interpretation critics tend to point out that the observer is outside the system of observation. But this is a weak objection. It is also a contradiction. Quantum theory exposes specifically that the observer is part of the system. This is because the scientific observer can only use scientific instruments and physical senses. We have discussed the consequences of this. Of course if we are validating the independence of a consciousness, which science does not do, then a level of consciousness can be used and must be (as recognised by Einstein that consciousness must be higher than that which it is evaluating). This is the very thing that the experimental set-up is not. As we have discussed thoroughly, the scientific observer (consisting of physical senses and scientific instruments) is part of the system and is inside the context of the experimental set-up. The fractal hierarchy shows that the 'outside' can only be in a relative sense. At the highest level there is no outside, but the absolute automatically gives a level higher than any of its creations (frequency patterns with a portion of itself in them). However, the scientist may still insist that the observed may be a quantum observation but the human observer a classical macro-observer (a technically finer point). However, when we take into account fractal levels of different orders and that basic consciousness (which has inherent quantum-computer properties) is part of everything, it isn't a consideration. Einstein also couldn't accept the idea of quantum uncertainty; that no matter how accurate one could be in the measurement there will always be an uncertainty in the result; that there is only probability.

Nothing is ever certain in the quantum world. For example, in between measurements it is meaningless to ask what is the position or velocity of, say, an electron or, say, the time at which an atom emits light quanta; they are entirely random. Hence Einstein's expression: 'God does not play dice'.

Again Einstein takes too small a view of the total universe and events of existence. Each individual projects his or her own reality (quantum wave pattern) combined with the collective effects of agreements, which establishes the selection of the common universe of apparent objectivity. We don't have to believe there is no form to the universe/environment when no observation/measurement is made. It may be a more blurred picture but even then this is under the influence of the next and higher aspects of the fractal hierarchy (or consciousness). For example, take the uncertainty principle conjugate pair, energy and time, in which it is found that both energy and time cannot be known accurately at the same time. However, higher dimensionally, energy and time can be merged into a single quantum state, and in fact in science it is known as quantum action. A quantum of energy will have a degree of extension in time. When we try to measure energy or time we quantum reduce the quantum action state. If we are measuring energy this will be separated out, leaving time virtually non-existent; and vice versa.

More recently Hugh Everett's many worlds interpretation has attracted increasing interest since it avoids quantum mechanics' lack of an objective universe. It proposes that all possible actualities exist in the universe, thus eliminating the collapse of the wave function to one possibility in providing the single outcome.

The quantum reduction, however, that we describe here does not suffer from this vague quantum-theory selection system in the collapse of the waves. It simply and naturally selects a reality by interaction and resonance of frequency patterns.

Everett's theory nevertheless doesn't deny any consciousness interaction even if we still wished to eliminate the 'magic' of the Copenhagen Interpretation. Consciousness, or whatever energy patterns scientists wish to believe in, can still do the selecting of a particular world view. The selection of a reality can still be governed by the individual's own reality. However, Everett's theory is interpreted as basically materialistic with a consciousness that is an effect.

Fortunately we don't have to become embroiled in an argument about whether the answer is Everett's many worlds or quantum physics' infinite possibilities of the quantum realm. The fractal holographic model solves this in presenting an objective multidimensional structure of different reality levels. We can therefore safely continue with our view of the Copenhagen Interpretation.

35.

SECONDARY QUANTUM REDUCTIONS

It is not being recognised that even following the initial collapse of the wave packet from coherence within the quantum realm to decoherence and selection of a single possibility in our material world, there exist many remaining coherent states (second-order quantum states). These are unfortunately being quantum-reduced further (unknowingly and sometimes knowingly) with complete loss of higher-order information.

As quantum theory became more expanded, the term quantum reduction developed two forms: objective reduction and subjective reduction. Professor Roger Penrose used the term objective reduction to mean that when the interaction occurred between the observer and the quantum realm, the collapse of the wave caused the manifestation of consciousness. This aligns with the view of many scientists, in particular, biologists, that consciousness is a by-product of the brain—such as possibly the field system around the brain and is the mind phenomenon. Other scientists believe, however, that the process is subjective reduction, meaning that consciousness was already there and this brought about the quantum reduction.

An initial difficulty for the proponents of the subjective reduction is that collapse of the wave is also caused by certain environmental interactions, which are not regarded as living things and therefore not possessing consciousness. Thus from the point of view of those that favoured the objective interpretation this would support their choice of objective reduction.

Although the answer to this, narrowed down still further the degree of support for subjective reductionists, nevertheless, there is an expanding influence within the peripheral sciences, New Age and spiritual philosophies, revealing that all energy is fundamentally of the nature of consciousness, and thus the subjective-reduction viewpoint can be restored.

To validate the subjective-reduction viewpoint we must have a consciousness that is or carries energy. We have indicated that 3D consciousness is a portion of the infinite potentia, the nonquantifiable absolute (no energy, particles, waves, at this level). But this is a 3D consciousness 'moulded' by frequency/wave-pattern templates, with which it (consciousness) is holistically coherent (the particle, wave). That is, pure consciousness (potential energy) is merged with and delineated by wave patterns (active energy) constituting this 'mind' consciousness. This 3D consciousness would originally be of the higher-frequency strata in the higher-fractal levels of consciousness, then projected down.

It would appear there are different types of secondary orders of the collapse of the wave function. We have previously mentioned the second-order quantum reduction of the falling body, which is brought out of coherence with space-time by a physical action upon it. A further example is the learning pattern in skills when just making simple learned physiological movements. It is in a superposition condition before it quantum reduces to 3D values for practical execution of movements (though the term 'superimposition' might be better in this case). For example, in a skill involving pressing keys, or playing notes on a keyboard in which finger action and whole arm are required, the complex wave can be imagined at the shoulder level that immediately quantum reduces to the appropriate lower fractal levels, such as the fingers to provide, say, one finger action after another. We shall say more about this.

For conscious control over movements one must be able to look ahead—one obviously has to know where one is going. But with physical movements this can't be intellectual; it must be information tied into the learning pattern. In fact it is the kinaesthetic sense which gives the future positions. This is the sensation of the programmes within the learning pattern giving a feeling in the muscles as to which direction to go in. This is also skill memory. Thus for control the computer system must look ahead, not the intellect which would cause the wave to collapse out of step, and we shall see that instantaneous control is necessary, indicating that the brain can't handle this—is far too slow—and we must postulate a mind that operates on quantum (instantaneous) states.

Now we have some familiarity and background information on the learning pattern we must examine in detail the simple parts involved in order to justify assumptions made. The reader may find difficulty in imagining reducing consciousness to such small elements, like a few bits of the computer, and then thinking from this viewpoint. It is too simple and that's what makes it difficult. Logic is the same; it is sometimes so obvious it isn't understood. We make too many unconscious assumptions, forming contexts (references). The following analogy may help to illustrate this difficulty.

Consider the thought experiment in which we have a person seated on a swivel chair. We imagine his mind has been cleared of all recordings and memories, and there is only raw, unmodulated consciousness—sheer awareness. He is in a fixed position looking ahead at object A (he can wear blinkers to make sure nothing else is seen). The requirements are that his mind is blank and cannot even think of moving—he can only see the object. This is his existence, his universe. As though in a hypnotic trance, he could be stuck there indefinitely—unless someone knocked his seat out of position.

Another individual now swivels the seat rapidly by 90 degrees so that the subject is now staring at another item, object B, and nothing else. Since no recordings are made, he is simply using a blank state of consciousness.

The key item now is that in order for him to be able to be aware of both objects A and B, he must not only have to record A and B (separate occasions) but must also record A and B, not merely stuck together (associated), but as a new state C; where C is not just the sum of A and B, but a new quantum state C, with A and B merged together as one whole. Note the resemblance to superposition in quantum theory—though this would be better described as superimposed. While the subject is thinking of A he can also hold in his mind, C (because A is in C), to enable him to be aware of B and thus be able to switch his viewpoint to B by going to C first (assuming he can turn to do this).

This nonlinear principle never seems to dawn within orthodox science, since science is engrossed in linearity and on the quantitative separation (ego) interpretation of external reality. And yet science knows that life, mind, and proper artificial intelligence can't work on program strings; linear arrays. How does this work in reality then? (Note that in the above, C, underlying A and B, constitutes a true nonlinear arrangement of two levels.)

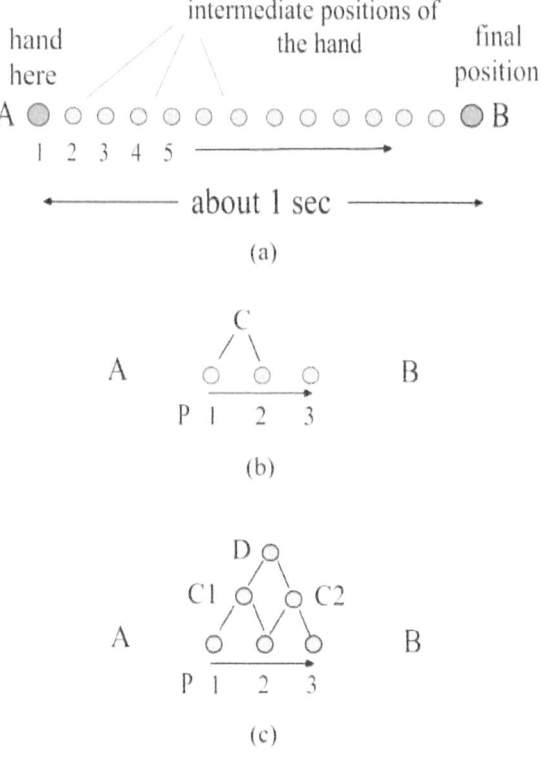

FIGURE 18

Consider the following linear string of computer bits (or units that can store one bit of information). If we begin imagining moving the hand from A to B, Figure 18(a), the learning pattern required is far too complex for our demonstration—nevertheless keep this in mind; it is only a difference in complexity. We can consider the separate bits 1, 2, 3 as producing or describing the separate positions of a point P as it moves to B. How does P jump from 1 to 2, etc.? Science calls it association (which is linear—a 3D connection side by side). Science only thinks in terms of the parts in linear relationship, sheer predictable machinery with no consciousness. The latter (consciousness) is a nonlinear mechanism itself (potentially infinitely nonlinear, the essence of our absolute, 'ocean' model). This belief (or erroneous scientific conclusion) enables the mechanistic view to have some possible validity in deterministic existence in which predetermined paths are triggered with no actual control or choice, just the illusion of it. However, science is now questioning linear strings of information (that is, rejecting them as inadequate), which leads us in the direction of a different type of control, resembling consciousness.

Thus computer scientists have recognised that linear strings of programming cannot ever explain the skills achieved by living things, even with the deterministic, predictable belief system. Science 'knows' that the answer lies in nonlinearity but refuses to recognise internal nonlinearity (inner-nested levels going into higher dimensions of integration—the essence of a fractal, holographic system). In psychology the stimulus-response connection—a particular response automatically following a particular stimulus—is not given a satisfactory scientific explanation, just 'association' or linear links.

Now look at Figure 18 again. For P to move from 1 to 2 we must apply the same principle as in the swivel chair analogy diagrammed in Figure 18(b). We add unit C as shown, which is a new (quantum) state and not 1 and 2 associated or 'stuck together'. The unit C will have a slightly higher frequency (see earlier section). Point P can now use C to get to the second position. But even with further C's in place, P can't get from 2 to 3. We must have a third level—Figure 18(c). Why can't we have three units (or more) integrated by one C? For the similar reason that in the extreme we wouldn't have one president of a company to, say, a thousand ground-floor workers. It would be very inefficient and there would be poor control. The learning pattern must have a fine gradient. We now see that to move from unit 1 to 2 to 3 with complete control we must have, in addition to the second-order nonlinearity denoted by C's, a third order denoted by D, Figure 18(c). This gives us a completed integration.

We also see here the beginnings of a new principle that the larger quantum state in going from 1-2 units to 1-2-3 units requires a position in the hierarchical levels higher up due to higher frequency. The principle is that the greater the wholeness (quantum state), such as consciousness, the higher the frequency. The greater whole is obviously the highest control and thus must have the highest, or higher, frequency (rate/density of information). Now we have some idea of this mechanism let us move to a slightly more advanced and practical example.

Notes

1. Book: Quantum. *The great debate between Einstein and Bohr* by Manjit Kumar.

PART SEVEN

COLLAPSE OF THE WAVE FUNCTION

The term 'collapse' might be considered a relative description; not a popular one as it stands. We have seen, however, that it is a little like Fourier analysis—a breakdown of a more complex wave into sub-packets of waves governed by the mode of interaction. Let us continue with the learning pattern.

36.

LEARNING-PATTERN APPLICATION

The collapse of the wave packet is even more profound than recognised by science. In addition to the primary collapse (Copenhagen Interpretation) there are secondary-wave reductions. These second-order collapses are commonplace phenomena in our everyday lives.

The reader may ask why are we expending so much attention to the learning pattern when we just want to know the mechanisms of quantum reduction (decoherence, collapse of the wave packet). The simple answer is that anyone can be familiar with the concept of a learning pattern, which is a mechanism and program that directs muscular movements, for example, in skills in which the learning patterns are well developed (well 'grooved-in'). Nevertheless the learning pattern provides all the complexity we require: fractals, holographic characteristics, frequency patterns, different orders and the property of non-linearity.

Now, the quantum realm (the state prior to quantum reduction) is probably infinitely nonlinear, though we need only consider a few dimensions and fractal-point manifestations in this potentially infinite gradient between the quantum realm and life as we know it. This is not as bad as it sounds. In simple analogous terms, compare it with the company organisation given earlier and many times in the author's writings. Thus the quantum realm is a little like the pyramidal ranking system of a company organisation, but imagine many levels of rank—in fact a theoretical infinite number, forming almost a smooth gradient. As a tentative illustration we could consider the bottom level, the ground-floor, sending a question to an upper level and receiving the quantum-reduced information (waves) governed or regulated by the ground-floor context (or even information transmitting between levels if one is higher up).

Thus the learning pattern could be the ideal candidate for describing a common and continuous example of the phenomenon of quantum reduction. Nevertheless one may object to the assumption or suggestion we have made that the quantum realm is infinitely nonlinear. Let us first remember that the quantum realm describes a mathematical state and is not taken literary by all scientists. The infinite probabilities and possibilities of the quantum realm can be considered in a state of infinite potential, and are not actual. It is basically our previously-mentioned absolute eternal potentia: nonquantifiable and beyond mass or energy. Nevertheless we shall extend this to include many energy states, which are simply not detected by current science and thus fall into the same category. This is the fractal matrix,[1] and the fractal, holographic orders of universe levels, that is, dimensional steps for existence and evolution—the exploration of and by consciousness.

We show again in Figure 19 the complex wave packet (or wave function) to show the reader this is still relevant. However, even this is too complex to begin a detailed explanation of the collapse of the wave and, moreover, to describe a computer read/write system.

Let us begin with a reference to a statement in an article Morphic Resonance and Quantum Biology by William Brown, molecular biologist at Hawaii University, in discussing biomolecular quantum communication and brain-imaging, that memories are created anew in the required moment.[2] This is similar to the activation of a learning pattern in physical skills (and note that the program in the learning pattern for learned movements is the memory). We shall shortly use the learning pattern as our model for describing quantum reduction.

Thus we are talking about a condition that didn't already exist as such prior to its selection—the selection itself changed the condition. For example, in detection of a particle, it is not in

The Emerging New Science

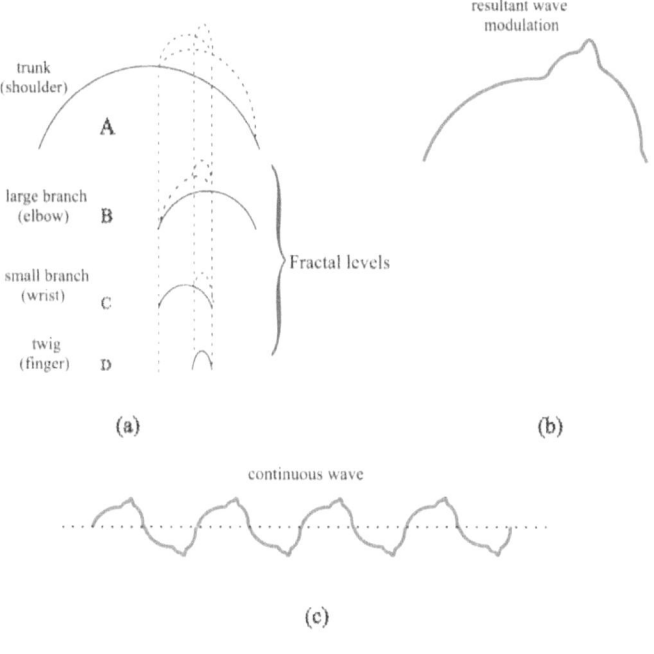

FIGURE 19

the localised particle condition prior to observation or measurement. The latter are all contextual: What is observed is in the context of the detecting system. Recall our example of the falling body, which is in a state of coherence with space-time prior to an action on it, such as interfering with its motion, including it striking and resting on the ground. It would appear that such second-degree or second-order quantum reductions are more common than we might think. In fact we are proposing theoretically countless levels or orders for the collapse or decoherence phenomenon. In fact decoherence of the wave packet will be continuous on different dimensional scales. This will take place between the fractals levels we discussed earlier. However, we need a suitable example that is complex enough to provide us with a model for a required generalisation of these extraordinary quantum features.

The learning pattern satisfies these requirements: a programmed mechanism directing all muscular movements within the biology of living organisms.[3] We shall choose the simple movement of the human arm and study how the learning pattern is developed, and that it is a four-dimensional holographic template that houses programmes and converts nonlinear information into linear information. The main feature of interest is the 'collapse' of the nonlinear complex wave into the 3D space-time elements to provide linear motion, one bit at a time (meaning small sections of motion or units of computer bits).

Figure 20 can represent a schematic for a multidimensional matrix, showing holographic connections that would provide the basic information capacity for these movements of, say, a five-finger exercise (or anything else appropriate) on a keyboard. It is the basic matrix of a learning pattern and is fairly ideal as a model giving a fine gradient. One unit at A, 2 at level B, 3 at level C, and so on. In practice, of course, one should imagine countless units.

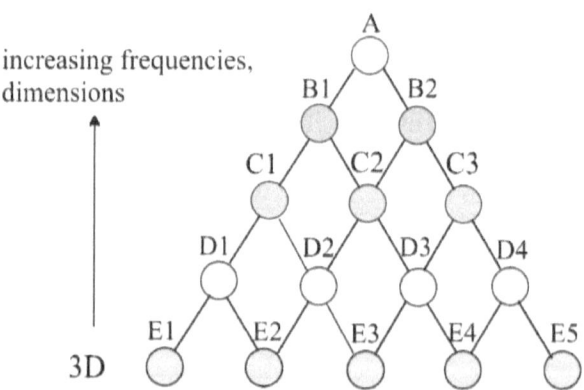

FIGURE 20: Basic matrix for programmes and construction.

An important point is that one knows or feels (kinaesthetically) the next movement, not by linear stimulus-response but by triggering the upper (inner) levels of the learning pattern towards greater unity, where the unification of the first and second movements is stored by means of a single oscillation, expanding up to more movements as a single oscillation at A. One can imagine in this highly simplified example that E1 to E5 each provide the energy (for changing the angle of the joints) for each of the fingers for playing the five notes. But realise whole 'pyramids' of information, such as this, extend downwards for each finger—that is, there is still a pyramid structure for even a fraction of one finger movement—down to 'bit' size. All the circles or letters represent whole quantum states—but could be broken down further.

Now as the first-finger movement occurs, E1 is focussed upon, and it passes its information up the hierarchy to the unity (D1) of both finger-1 (E1) and finger-2 (E2). The state D1 is not only the resultant of E1 and E2 but is in higher-dimensional format (for example, say, 3.1 dimensions and higher frequencies). Unit E1 will resonate with its own counterpart in D1 causing a breakdown, a quantum reduction (compare Fourier analysis) of D1, releasing wave pulse E2. Unit D1 simultaneously resonates with C1, releasing D2 to give a superimposed feeling of the next position for E3Cbut more 'distant' or faint. And further, C1 resonates with B1 to release C2, giving the superimposed sensation of E4 through D3. All this is simultaneous except that timing intervals may or may not be programmed, governed by the intention (for example, for consecutively played notes, or playing notes simultaneously).

Let us give the reader a background on what the learning pattern is capable of, though it will be a simplified version (for instance, not one with the true holographic properties). To explain the property of the learning pattern regarding its integration aspects in time (as well as space) we must consider a completed learning-pattern template. That is, the movement has been learned, say, of the arm and hand, from A to B (if it is a simple straight movement then anyone has already learned this without any specific practice). See Figure 21.

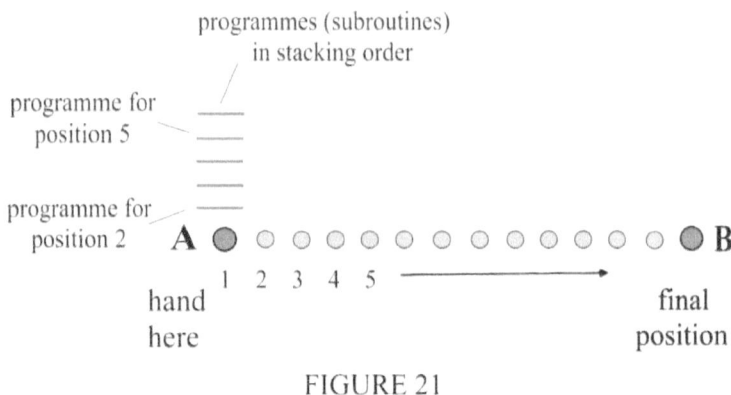

FIGURE 21

If one studies Figure 21, one sees it is showing that the program for the different positions of the hand into the future, for about one second, are in stacking order when the hand is in position-1. (Note that we have an analogy with our computers, which utilises stacking orders.) This means when the hand is at A, in position 1 (but intending to move to B), it not only has the programmed coordinates for position-1 but all the future ones, in order, first to last. In this example, it means, say, a thousand separate pieces of information corresponding to the exact positions between A and B are evident within the kinaesthetic sense at A, sensed in the muscular tensions acting in the margin of consciousness. However, all these coordinate frames are perceived as one whole, instantaneously, with potentially instantaneous breakdown—this is a holographic coherent structure of superimposed quantum states.

Strictly speaking it is not quite as simple as depicted in the diagram, for example, these future positions, such as the final one at B, are actually whole quantum states—between the beginning, to the respective separate positions. This is an example of quantum superposition and superimposition. These quantum states (wave patterns) mould the motion, pulling it to the position at B. Each one individually superimposed. [A well-developed learning pattern can handle about one second of data simultaneously and collapse this coherence from its nonlinear condition to linear for application in 3D space and time, such as one finger at a time, but this time interval will vary according to the skill.]

We mentioned 'stacking order' above. This is an example of coherent quantum states in superimposition, that is 'vertically' with increasing higher orders, and can be sensed all at once. Thus this is superimposition in the same time, until quantum reduced into 3D spatial and temporal linear sequences. The order sequence is triggered at the 3D level by the first motion (for example, a finger). Each state contains within it the information of all the future ones within the span/integration (say, one second in time) of the coherent quantum state at that moment. The physical condition triggers the next movement in the sequence like falling dominoes. As the stacking levels reduce, the top ones are replaced by future ones extending a constant reach in time, until the end of the sequence.

These are secondary quantum reductions compared with the known primary ones of quantum theory. The mind can hold these coherent simultaneous states but the environment can't (recall the section on the requirements of the quantum computer).

All this occurs quite automatically but the conscious mind can interfere. The superimposed coherent multi-quantum states in the learning pattern mustn't be looked at, such as thinking of a state too high in the stacking order, corresponding to a new future position (which is not the next one). One must let the stacking order feed back its information and order to consciousness one state at a time—this is the 'letting go' sensation that one experiences when a skill is learned. If the conscious mind (intellect) takes over control, it will collapse the stack in wrong order.

The wave function, however, provides the necessary requirements. In terms of the collapse of the wave it means this complex wave packet is initiated (that is, we are ready to make a movement) and comprises all information (modulated sub-waves) on the future positions and sub-positions to point B. This is all at once. But clearly it has to be broken down—collapsed in an ordered manner, governed by programmed instructions, and applied throughout the interval of time (about one second). Note that one second is a long time in terms of computer information, and quite a high level of skill is required to provide this. However, everyone with normal coordination has such a learning pattern for the simple straight motion of the hand from A to B (any distance, any time interval). How can anyone know that? Recall all those times (or try it now) when reaching for an object, in particular, at the meal table, say, the salt pot about 18 inches away. One first glances at the object, and during that moment one can (or could with trained perception of the kinaesthetic sense) feel kinaesthetically the hand grasping the salt (before moving) instantly as one estimated the distance. We can now even turn away and the hand blindly will go accurately to that position and grasp the salt pot.

Thus in Figure 21 the movement from 1 to 2, or 1 to 3, or 1 to 4, etc., are whole states, and now we see the holographic property more clearly: the whole movement from A to B is within or connected to all the sub-parts (1 to 5, 1 to 4, etc.) and the initial starting point. This is the whole-within-the-part feature of the holograph. Each state only gives a pull towards its own respective position but when they are all superimposed they delineate the precise geometric path learned—extending into the immediate 'future'. This wave function complex translates into spatial and temporal position for all movements.

All these frames (1 - 2, 1 - 3, 1 - 4, etc.) are sensed kinaesthetically (a sensation seemingly in the muscles) by the individual. This is why miraculous skills can be executed and be controlled at every point in space and time. This is what holographic means—the whole within each of its sub-parts. It incorporates both quantum superposition ('horizontal' interconnectedness) and quantum superimposition ('vertical' interconnectedness). As the hand moves to position 2, this position now becomes the new position 1, the first point. The final point at B moves ahead one step (if the sequence continues). There is a kind of scrolling motion.

Let's use a computer analogy. Picture a computer screen filled with text. Note that as we scroll down, the top line goes up and disappears. The bottom line, in moving up, has left a space for a new line to appear. This process continues as we scroll. Imagine the computer screen of information (text) is the span of the learning pattern. In going from A to B, position-1 would be the top line, and one only reads the top as it scrolls but can, in the case of the learning pattern, instantaneously scan ahead (to the bottom of the screen—compare sensing the whole learning-pattern information from A to B).

The learning pattern quantum-reduces from a nonlinear state to linear. Thus learned sequences of movements do not have linear strings—which scientists have long since suspected (that they do not). They have information stored in a nonlinear state: the complex wave (Figure 12 or 19), wave packet or wave function that contains all the sub-levels of information down to the highest differentiation of parts inherent within the learning-pattern development. Let us elaborate on this.

For example, in a learned sequence of movements at the piano, the action of a single finger with its linear pulses (compare bottom line in Figure 12) causes its (the finger) simple wave pattern to resonate with the larger complex learning-pattern wave, say, at the top, which has the wave forms for wrists, elbow and shoulder (piano-playing must be a whole-arm action). The resonance causes a breakdown of the complex wave as its own components, of the same frequency, are resonated/activated, selecting the next in stacking-order pulse(s) (in time) for the next finger action. Of course, it is far more complex than this since the finger waveforms also—and simultaneously—select from the wrist and elbow wave forms. All these higher fractal levels quantum reduce to provide the wave form for the next finger action (or whatever). Of course, the higher-wave packet remains intact—it instantly quantum regenerates itself.

Thus the fractal system of higher orders exists within the learning pattern computer system of the mind in operating physical movements. This hierarchy from fingers to wrist, elbow and shoulder was experientially verified by experimental psychologists more than 30 years ago. (The experiment was described in one of the text books used in the author's experimental psychology doctorate course—reference lost.)[4] Let us explain this.

Briefly, subjects learned movement sequences involving pressing keys on a type of keypad. However, ingeniously their joints, from fingers, wrist, elbow, to shoulder were restricted appropriately. They discovered that if the shoulder level only was left free (elbow, wrist, fingers were bound), when a sequence of movements pressing keys was learned, this learning information would pass down—what the psychologists called the ranking system—to any of the lower joints. The latter could execute the pattern of movements as though learned themselves at any of these lower joints, when they were freed, say, one at a time. For example, in freeing, say, the wrist (with fingers, elbow and shoulder bound), the wrist now used the learned ability (learned at the shoulder level). However, the investigators further found that when a lower joint level was freed and it learned a pattern, with higher joints bound, the learning did not pass up the rank to the upper joints.

Thus learning would pass down but not up, exactly what one would expect from Figure 12. A higher carrier wave can quantum reduce its information to a lower one, but a lower level can't pass any learning up to an upper level.* Looking at this another way, an upper-level joint learning pattern can 'write' to a lower one, but a lower one can only 'read' an upper one. This means the finger level pattern exists also at the wrist level (coherent with the wrist carrier wave), and also at the elbow, then shoulder level. The shoulder learning pattern would have the finger, wrist, elbow patterns coherent with the highest ranking wave function at the shoulder joint level. Of course, the experimenters didn't have the theory for this. [* One cannot obtain higher orders from lower orders (such as in the Big Bang theory) except in quantum regeneration in which the dormant or virtual higher intelligence already exists.]

Take a look again at the read/write (computer) system in the last paragraph. This is the essence of quantum regeneration and quantum reduction. Life, consciousness, is doing this continuously. In fact, this is precisely what true evolution is, described today often as ascension. However, applying this to human evolution, the lower level (for example, human 3D consciousness) needs to select not just its own level of reality from the upper-wave packet as with the learning-pattern example above, but to quantum regenerate the lower level to resonate with a higher level so the latter doesn't collapse, that is, quantum reduce to the lower level. This is achieved by striving to improve his/her qualitative behaviour (self-improvement), and a 'read' is not merely made to the higher level. A write process occurs in which the higher level 'writes' to the lower level (a new state for the lower level). Note that it also means the lower level 'writes' to the higher level by bringing in additional wave forms from the lower level and ordering them (quantum regenerating). Thus when the selection is made from the higher level and quantum reduction occurs, normally to the same initial lower level, the individual's inner/higher levels of consciousness can be accessed using the intuition, enabling a higher state (frequency pattern) to be drawn down (and they move 'up' on a long-term basis). In other words, the individual will draw in higher orders of frequency patterns and gradually 'ascend'.

The learning-pattern examples we have given have all involved learned programmes. Let us briefly explain what happens during the learning process.

We wish to learn skilfully a particular movement and thus practice it—which involves repetition of that movement. There is a very fine relationship between the input (conscious mind) and the learning-pattern programme. Both are or have frequency patterns, which will be in resonance depending on how well learned the movement is. Most movements, such as walking about, are well learned and the program and input consciousness form a holistic coherent system (one whole), which causes consciousness to think that the input itself is the learning pattern. (The author practised introspection and learned to separate these two—that is, identify the self-component (conscious-mind input) and perceive the mechanisms of the learning patterns as separate.)[5]

In learning a movement (a skill or part of a skill) the input (conscious mind) and learning pattern (mechanism) are in a subtle cause-and-effect relationship. The input, as cause, directs the learning pattern, and the learning pattern obeys but instantly makes a copy of the input's efforts. In effect, it forms traces of a 'groove' (remember it is geometric). Now when the input tries that same movement again, the learning pattern intermittently acts as cause and feeds back the 'groove' to the input to help guide it. The input, with this help, does a better job the next time, and subsequently the learning pattern makes the copy again, a better one, and feeds this back. The input and learning pattern are working in unison to build up an accurate programme. As the learning-pattern program becomes well developed, the input 'let's go' more, that is, allows the learning pattern to be cause and the input to be effect.

In very high skills the cause factor from the learning pattern becomes more dominant, and with the reducing effort and increased relaxation of the input, a condition may be set up in which the learning pattern (which potentially can be expanded indefinitely) turns into a habit pattern and no further basic improvement occurs, merely a final consolidation of the existing programme.

When age is also added we confuse this plateau of learning ability (then followed by actual deterioration) as due to aging. But it can be avoided, or the learning pattern can be restored to its natural flexibility. Physiological deterioration is inevitable with age but this is quite different. However, we don't need to pursue this further in this book.

The procedure adopted in acquiring information on the learning pattern was mainly the practice of mediation and introspection. It is possible to identify the self-aspect from the machinery the self uses. A very simple example that requires no training is for anyone to picture in their mind a memory. They are then looking at a mental image. They are not the image; they are not inside it or in its space. They are looking at a mind recording; a frequency pattern, an energy structure. All of the mind is a structure but with an operator (basic consciousness, sentience) merged with the machinery of the mind. The sentient part can separate from the recordings but it can't separate any more factors from itself—that is, can't separate from itself, thus identifying a more basic self than, 1) the body (including brain, mind and consciousness), or 2) brain (including mind and consciousness), or 3) mind (including consciousness), and finally 4) consciousness, the input self, a single whole unit (a projection from the higher aspect, or fractal level that we call the soul).

The input will have a complexity of higher frequency patterns moulded into it from higher fractal levels but uses the mind to step down (transduce) frequencies for the necessary interfacing with the central nervous system and body. By increasing the awareness of the kinaesthetic sense it is possible to identify oneself as the basic input to the learning patterns and precisely how one, say, keeps the machinery of the learning patterns going in executing movements. In particular, to know how the conscious self controls these automatic mechanisms, allowing them to run the body machinery (voluntarily) and be in complete control every split second, and yet require little to virtually no thinking other than intending a result.

Compare the imaginary situation and analogy of the president of a company (at the top of the ranking pyramid) who knows simultaneously overall what is going on in the company and in fact is merged with all the components (manager, ground-floor workers, etc.) but can vary the focus in any region to change or aid that region. If all is well (programmes thoroughly learned) the president merely monitors the activity of the company, but it will stop, immediately his or her attention is withdrawn. The reducing ranking system (executives, managers, foremen, etc.,) literally amplifies out the details. The single ideation of the president is subtly transduced (broken down) into quantitative components.

The input energy is precisely the amount of attention that is absorbed/required by the learning pattern to execute the movements. As the learning pattern is expanded (holographic amplification), the input (attention energy or consciousness) is 'diluted' but spread more over the unified time interval, mentioned earlier—the time span of the learning pattern, Figure 18 refers. As the learning pattern expands its input/output ratio decreases (indefinitely)—the 'president' hires more 'staff' to multiply his efforts.[6]

Thus it is possible to use mental empiricism to locate and identify the self-aspect and view the body, brain and mind as mechanisms.

37.

CREATING (SELECTING) ONE'S OWN REALITY?

> *Scientists, in researching the most important feature of quantum physics, the collapse of the wave packet (primary quantum reduction), unknowingly and continuously, in secondary quantum reductions, collapse the specific data that is being observed and destroys the solution that is sought or could be revealed.*

We shall now continue using the learning-pattern example where appropriate as we explore the phenomenon of quantum reduction and selection of one's reality. Note that in the expression 'creating one's own reality', the self is much more than just the conscious mind and includes the total self, for example, the unconsciousness. Let us now remind the reader there is not such an abrupt change from the quantum realm of quantum theory and the manifest material reality, as implied within quantum theory (incorrectly). There is the immensely complex fractal spectrum of dimensions and subharmonics. All levels can quantum reduce to a lower level; that means the 'mysterious' quantum state is quite 'tangible' at its own level (recall the great scientific debate between Einstein and Bohr).

Before continuing with the learning pattern in demonstrating the reduction of the wave packet, let us remind the reader of the dominant role of physics and, in particular, quantum physics. That is, the basic science is physics; all disciplines of science ultimately become physics as they advance. We have then stated that physics is the study of how energy works—how mechanisms function. Here we shall point

The Emerging New Science

out that quantum theory is the most successful and advanced subject within physics. What might be considered to be the most important concept? The founder of the wave equation, Schrodinger, once said that non-locality was the most important feature. We are not disagreeing here but are recognising (as does quantum mechanics today) that nonlocality is a part of the broader principle of collapse of the wave function or quantum reduction (which we have explained).

Thus we have arrived at quantum reduction as being the most important phenomena, at least for science to evaluate (the consequences of this are the solutions to the nature of consciousness, life after death, religion and origin of the universe). In terms of the layman's language, this is the same subject, reiterated over and over again throughout the New-Age movement, and beyond, described by the statement: 'creating one's own reality.'

We can now use Figure 12 to represent a learned short sequence of movements of the whole arm, including elbow, wrist, and fingers. That means the learning pattern is fully coordinated in its bits (note that this doesn't mean this learning pattern can't increase its number of bits, raising the potential for coordination/skill).

Note that it would be too complex to give more detail to this configuration and the reader must envisage the general idea of how these learning patterns work. The complexity may be realised when we point out that each part within the whole has the same fractal form (of bits). For example, one finger will have an organisation of bits in this fractal pyramidal relation; use Figure 19 if it helps. Moreover, these learning-pattern pyramids of bit/waves will link together amongst the fingers, forming a larger learning-pattern pyramid configuration; similarly for wrist and elbow (though of course there is only one wrist, elbow, or shoulder). And then these are all linked into larger learning patterns; the complex wave in Figure 12.

The wave packet at the shoulder level has all the bits or elements of information of the lower levels, linked up to form a holistic system. This means the whole waveform, for example, a movement of arm, hand, individual fingers over a short distance of the keyboard, spanned by the learning pattern, let's call it A, can be controlled or accessed from any point within it*. By moving the first finger action, a 'read' pulse impinges on learning pattern A, and the whole wave packet for the A movement is momentarily quantum reduced from the shoulder level, where full coordination of all fractal levels is taking place. The whole state of A is immediately reduced to its appropriate parts in space and time at the finger level (contact with the keyboard) —and at the wrist/elbow, accordingly—to provide normal 3D performance. Thus this action of the first finger begins the conversion of the nonlinear learning pattern into separate linear movements. [* 'any point' means any coherent group of 'bits' that are acting independently, that is, as a whole; and are thus equivalent to one larger bit.]

Science, in its desperation to maintain non-coherent force systems (meaning that the only reality is materialism), readily renders the features of the quantum (holistic) realm as mathematical and not part of everyday reality. Aiding this confusion is science's refusal to model or even recognise true unity (yet it is all right to have it in the quantum realm—since its reality (unity) is dubious—as per current science), and it strangely has the habit of disappearing when observed/measured.

The fractal holographic system gives us massive superposition and superimposition, even quantitatively in the 'quantum realm' of lower orders. Recall the example of the falling body in a lower (second order) quantum state, but quantum reduces when acted upon, losing its coherence with space-time.

In the learning-pattern operation, input consciousness has all its own potential ideations (such as the idea of moving from A to B), which we can call action concepts, duplicated (copied and stored) by the learning pattern itself[7] (action concepts). The action concept is any state of attention/awareness on any movement or part of a movement or combination of movements during, say, a complex coordinated sequence. Each of these states, theoretically infinite in number, is a whole quantum state. The learned pattern is kept activated by these action concepts, that is, ideations, which can be quite vague, but the intention must be precise. The learning pattern continues to record all movements precisely; the computer system is making a quantitative copy of them (since merely an idea of making the

The Emerging New Science

movement from A to B without using the program will not give a high level of skill). This mechanism enables the input (the conscious thought, intention) to interface with the learning pattern system at any ideation level. This is the nature of holistic systems; activation can be maintained from any part (such as one finger or a wrist motion) ranging to the whole.[8] All frequencies are in harmonic balance in a learned sequence.

Scientists have not been recognising the extent to which that 'mysterious' coherence of the quantum world exists in our real world. For instance, in the falling-body example, when it collides with a body, such as Earth, it cannot continue with the gravity flow, and the gravitational force and Newton's laws immediately act on it. Thus coherence will be preserved only if the body is not acted on by a Newtonian force with a two-to-three dimensional contact; it is then a one-bodied system with space-time. This is the same principle for a body in orbit, for example, Earth in orbit around the Sun, which is also in 'free fall'. Such examples are coherent group states, present with random states, whereas the first-order quantum realm, also coherent, does not have any separation.

These examples have not been recognised. They can be regarded as of second-order reduction/decoherence. However, science has made some progress in identifying systems that remain in a coherent state while in 3D space-time, such as the Bose-Einstein condensate, as mentioned earlier, or the phase-entanglement of twin particles at any distance apart.

One might see why science is not detecting these more natural phenomena of creation, which can form the basis of a harmonic science, as opposed to our non-harmonic science and technologies (to which we have already given attention). The reason is that science does not recognise the nature of unity, consciousness, true qualitativeness, and that laws such as Newton's can be bypassed. As stated, even a spacecraft propulsion system (Figure 5) is functioning in one of these harmonic coherent states. Thus we can now see one of the factors why science has somehow skipped harmonic science and technology (in addition to a further powerful subversive 'political' control over energy development).

The current close-mindedness in the scientific community, the programmed belief systems that there are no perpetual energy sources, and the laws of physics, such as Newton's or the conservation of energy, are considered universal and must be obeyed, and this has prevented the recognition of these second-order coherent states all around us. All essentially brought on by the particular mode of observation that science makes, destroying these states, including when evaluating extrasensory perception. A coherent state has naturally the characteristics of superposition/simultaneity and its parts, in harmony, automatically quantum regenerate the collective holographic fractal structure. (Recall that nature, creation, and the universe begin with a whole quantum state, which then quantum reduces to a lower reality, but life on the lower level can quantum regenerate to the higher levels.)

Coherence is essential for most paranormal observations to be effective. But collapse of the wave is continually giving the observer (or apparatus) source the reality that it already puts out. Thus science's sceptical state will not encounter coherent phenomena—Einstein was correct when he made the statement: What one can't imagine one can't discover (the state of mind determines what it will see). This is also one of the reasons why science is quantitative and not qualitative; it frowns upon the state of intuition, imagination and emotion, and avoids them. Its methodology quantum reduces these more coherent states of energy, destroying the higher order, making them appear insubstantial and vague.

The learning pattern is also an ideal example for illustrating the gradient of reducing quantum states of many energy levels, rather than the abrupt observations of the current model of the collapsed wave from the quantum realm to the material reality. The learning pattern mimics the fractal system of the holographic universe and has every quantum-state level from total coherence (unity and superposition) for learned movements, to the smallest individual 'bit' pattern that functions independently.

This is why science doesn't recognise (discover or detect) holistic systems. As we have stated over and over, science is quantitative; the procedure in scientific investigation is analysis, and the mental attitude, which means the background context, is one of particles stuck together by forces. This is the intellectual state of mind devoid of the right-brain characteristics, such as intuition.

This state of mind creates a logical procedure that only accepts the part as being dominant. If scientific procedures examined a holistic system, the system would immediately lose its coherence before any knowledge about the system was gained.

If an individual is looking at a good work of art and is appreciating it, the moment he or she examines the work of art logically, that is, involving a complete re-focussing of consciousness from the whole to the parts, the perception of coherence is immediately destroyed. Note that the 'coherence' is the conceptual ideation of the art (the meaning) in the mind of the beholder. Thus this type of quantum reduction does not involve physical changes in the universe. In this art example, it is a response (in the mind) to what's on the canvass (in the universe).

What we are proposing here is that our (science's) nonqualitative approach to the acquisition of scientific knowledge, plus the accompaniment contextual mental structures that form, are automatically collapsing the 'wave packet' to the 'reality' level of the observer and the scientific instruments. Any state of mind will select that reality level, which is governed by a spectrum of wave patterns. The source of a quantum interaction selects or perceives its own level of data—hence the ease with which humans fall into arguments. Science creates its own belief system and selects only data that resonates with the belief context.

Although the examples of quantum reduction and decoherence appear widely different, this is only at the external material level. The behaviour of the frequency patterns underlying this would be expected to be considered the same—given by the structure in Figure 12.

Thus there is nothing mystical about the collapse of the wave function from some untouchable mathematical reality to observable reality. It is part of the fractal spectrum of frequencies underlying creation (Figure 1) with a range of wave patterns that interact up and down these levels and undergo transduction of frequency values. This is a frequency down-stepping and up-stepping process over the range of this fractal hierarchy. A lower level may 1) select from a higher level but only within the possibilities of its existing accrued or acquired frequency patterns, quantum reducing the higher level to the lower level (interpret in 3D but retaining the higher coherence) to bring understanding of it at that lower level, such as, bringing in great musical concepts to then apply quantitative/intellectual methods (write out the composition), and 2) the lower level (other people) can receive through sound the composition and then by quantum regeneration unify the parts and step up the frequencies, which then draws the lower state up to the higher one, and (with appropriate musical ability) achieve coherence and integration of the higher-level musical concepts and the subsequent understanding. In the above, (1) and (2) can work together. This is also the nature of true evolution.

The irony is that science is investigating the meaning of collapse of the wave function but in the process of doing so, it is continually collapsing the wave. As mentioned, science claims to test certain psychic, or religious phenomena, or even claimed unproven physics principles, and finds no evidence (even if valid) because the scientific procedure and apparatus quantum reduces the coherent states of resonance required, for example, in psychic phenomena; destroying all possibility of proof.

Thus as stated in the abstract, 'Scientists, in researching the most important feature of quantum physics, the collapse of the wave packet, unknowingly and continuously collapses the specific data that is being observed and destroys the solution that is sought'. However, the immediate response to this would be that they are fully aware of this theoretical quantum reduction while making measurements; this is the Copenhagen Interpretation of the well-known collapse of the wave. But we are talking about the second-order reductions.

Let us summarise examples of everyday quantum reduction of coherent cases: 1) The body in free fall is in a state of coherence, until 'measured'; 2) the 3D elements of programmes within learning patterns are phase-correlated in the holographic space-time template until a physiological move is made; 3)

The Emerging New Science

during art and music appreciation the human mind is in a quantum state, the coherence brought about by ingredients, paint/spatial, sound/time, until analysis is applied, such as a simple intellectual thought.

Scientists are being taught (are selecting) a reality level of knowledge, say, the belief in the immutability in Newton's laws, then this reality context within the mind attracts its own truth. This occurs when there is emphasis on left-brain activities, such as intellect, logic, and not right brain such as intuition. The intuition quantum state is of higher frequency and capable of tuning into higher orders of truth without collapsing the frequency pattern to a lower one.

There are different conditions of the quantum realm and its manifestations into 3D. Firstly we find that there are two reciprocal quantum processes: quantum regeneration and quantum reduction—already covered. Secondly, these manifest in different mechanisms: the laser; bodies in free fall; art; learning patterns; persons on same wavelength; advanced propulsion; uncertainty principle measurement; Bose-Einstein Condensate; quantum nonlocality and entanglement, and more. We see there are also collapsible quantum states not just in the environment but in the mind, emotions, mental and spiritual states.

Returning to the learning-pattern example, can we arrive at some idea of the hardware; clearly it is not likely to be pyramidal shaped, which indicated only software principles. However, we alluded to the fact that Figure 12 indicated the hardware. Recapping, the quantum states have a frequency and we saw the sine-wave representation in Figure 12. Now realise that a sine wave can be expressed as a circle, and if we translate Figure 12 we would have circles, preferably picture spheres (the real hardware). In fact, the sine waves can be representative of spinning spheres (higher-dimensional vortices).

Let us use the more complex diagram in Figure 20. We described this earlier. Each of the units, for example, E1, in Figure 19 are oscillations/waves. Wave A is the resultant wave (but can act as the controlling wave). Picture the resultant wave in Figure 12. This is more simple and we can clearly see the waves within/upon waves. Now the phase angle between them translates as distance. If we simply recognise without trying to visualise higher-dimensional geometry, the sine wave would translate into spheres within spheres now occupying 3D space. Thus we translate Figure 20 to look like that in Figure 22, which is now a more accurate hardware rendering. (The spheres are higher-dimensional vortices, recall Figure 4).[9] Each sphere in Figure 22 spirals inwards, branching off into smaller spheres and increasing in number. They are stable standing wave structures since each sphere contains within it two opposing spirals (vortex/antivortex). Recall Figure 3. As one may gather from the earlier figures on the vortex, this formation of spheres within spheres is fundamental to all complex structures, including the total universe. It is a fractal holographic system. The fractal system of a tree is an external projection from its basic holographic template of spheres within spheres in the tree's inner space.

From these diagrams and descriptions of the quantum processes we can see that nature (in this case, application to physical mobility) functions on geometrical information. For high levels of skill, that is, ability in moving and controlling the body, it is essential that an advanced reference system is used. Think of the robot arm in the automobile factory. As it reaches over the vehicle to do a specific job, imagine we shift its rigid installation a little, it would of course miss the target. Earlier robot arms only had one reference point. Compare someone playing a learned sequence at a keyboard. They can move their body and immediate corrections are made to allow for the movement so that the program continues unhindered.

This is the holographic learning pattern in space and time. By studying the system given here, of spheres within spheres, Figure 22, we see that all smaller spheres (vortices) are located with respect to, or in the context of, the next larger sphere. In this way the exact location is known of every independently-operating component. We may extend this to recognising that the 'bits' or particles can be precisely located and the location controlled—a necessary learning-pattern function.

The Emerging New Science

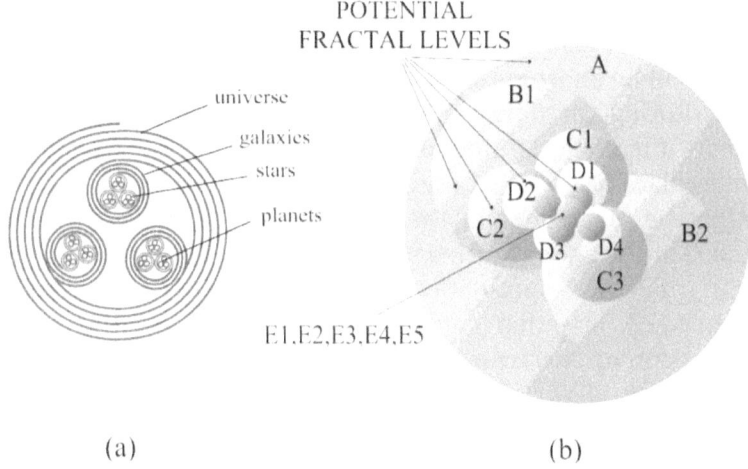

Basic fractal holographic configuration. Each sphere is a 4D to 3D spiralling vortex with its antivortex along a 4D direction. Note that (b) is a general template and not ordered specifically as in (a).

FIGURE 22

Now the structure in Figure 22, if projected geometrically (into a 2D diagrammatic representation), it will give us an idea of the wave-packet application. Each circle (sphere) is equivalent to a sine wave. Thus a summation or superimposition of sine waves, which acts as a single wave with a complex resultant (modulated sine wave), can be projected to produce the spheres within spheres. Compare Figure 12. The precise positioning of any sphere is thus given by the phase angle of these sine waves (this is one of the keys to how nature's computer system works). As we have stated previously, natures' mechanisms work on geometry, which is an absolute system. The universe operates on geometrical intelligence. Keep in mind Figure 4.

Note that if sphere A, that is, a single overall sphere is not present (learned—linking together the inner parts), then the learning pattern is not complete. The holographic extension in time will be short with poor linking together of movements and constituting an unlearned sequence—the sequence of movements it handles has not been learned. The attention would have to oscillate between B1 and B2 giving a sensation of effort. This might be for actions of the left arm (B1) and right arm (B2), which one is endeavouring to coordinate.

Further speculations regarding the configuration in Figure 22 are that information is stored in the (higher-dimensional) surface of the 4D spinning vortex in the form of sine waves of various phase angles when representing a (geometric) programme. Recall that this vortex has its opposite in the anti-universe/anti-particle side (coming out of the centre) and intersects with the one seen in the diagram. When the program is activated, it spirals towards the centre of the sphere, where intersections occur to create holographic forms.[10] For a universe vortex, 3D would be made up of the smallest spheres nearer the centre of the universe sphere. A galaxy would be a large sphere inside the universe, with its information stored in its surface and projected inwards to form stars, or sub-level information in learning patterns. Note that the surface of these vortices is higher, dimensionally, than their centres. This also applies to the learning-pattern structure, the tree, and the limb (e.g., arm). All this projects out schematically to form the pyramidal fractal dimensional hierarchy.

38.

CONCLUSIONS

Quantum physics, the great debate between Einstein and Bohr, evolution, intelligence, higher orders, all is contextual, there is no objective universe out there, such concepts are enough to deter the mainstream scientists let alone the common reader. However, if we put it another way and said that life, in interacting with its environment in the most fundamental sense (the quantum realm), attracted its own reality level, this might be recognised as being of great importance.

With no great knowledge of anything, most people have heard the comment, and understood it, that when science attempts to understand life, it dissects it and thus either destroys it or it is already dead. Something disappeared. However, we are not suggesting that all things are alive in this way but merely indicating that science has its own reality level—which is the quantitative particle level—and will either not detect the higher order, or higher coherence of the universe, due to insufficient resolution in the detection procedure, or will unwittingly quantum reduce it to a lower reality level (secondary reductions). Thus science destroys the higher order and truth (and higher scientific laws). In a nutshell, a lower-order mechanism cannot detect a higher order, or (depending on what it is) it reduces or selects from the more complex waveform the resonant order that matches its own, and the higher order or truth is lost.

Fortunately consciousness is real and is also fractalised; it already has higher aspects within itself which can be accessed by intuition so that it can not only match any level of coherence in the universe but be one level ahead (consciousness must be above, that is, outside the immediate context of that being evaluated—the scientific observer is on the same level; see previous sections).

This subject of creating one's own reality is almost like a joke or puzzle that man needs to wake up to or solve. There is no wonder we argue so much. Each person selects his or her own reality level. For example, when perceiving a higher state of (quantum) coherence, such as in the field of parapsychology one's own wave patterns will tend to collapse the coherent quantum state to the sub-harmonics of the state being observed, measured or interacted with.

Let us recap and summarise the main points of what we have introduced in quantum theory. Firstly there is the primary collapse of the wave/frequency pattern from the interaction with the quantum realm of 1) life sources, and 2) measuring devices and energy sources, generally. The quantum realm would be the higher order.

Secondly, further processes besides quantum reduction are quantum regeneration and their secondary orders. These interactions are clearly involved with creation in general. But this gives causality to individual sources to participate in creating their own evolution, or in fact their downfall. Thus an individual is drawing in frequency patterns governed by his or her identity (frequency spectra). If their interaction with the environment/people has been predominantly positive, their personality vibration will attract from the infinite possibilities the next higher layers of the fractal system of increasing higher orders, subsequently quantum regenerating higher quantum states. But they must put the higher order into the elements of existence (picture wave patterns in Figure 12), such as in behaviour, responsibility, ethics, benevolence (states which have harmonic frequency patterns since these energies cooperate with the environment). If their spectrum is higher than the environment they will tend to automatically quantum regenerate the incoming information, that is, put order into the lower level (quality into quantity) and thus draw in (select) a higher state within the quantum realm/potentia, ultimately raising the person to that level. Personal thoughts, actions creating frequency patterns of greater coherence and qualification, will quantum regenerate within self, higher-frequency patterns, which then select from the fractal spectrum a higher level of ascension. These are all mechanisms of existence that physics could handle.

Note the early terms of quantum theory that the quantum reduction occurs from a condition (in the quantum realm) of phase-correlation, to our material world of phase-randomisation. Scientific analysis is constantly reducing the 'correlation' to 'randomisation' instead of retaining secondary correlations (coherent states). We could say that the primary collapse creates the particle world, with particles, objects separate at the 3D level. But as we have indicated, many natural states—not artificial or technological—have groups/regions of particles in coherence (such as repeatedly described in the free-fall phenomenon). However, ours is an existence of a high degree of objectification and we interact by 'force'. That is, not only do we interact two-to-three dimensionally with matter (matter against matter) but that this procedure in artificial processes causes the secondary quantum reductions to occur, bringing about our familiar Newtonian decoherent condition (phase-randomisation/ quantum reduction). We might consider that we are obsessed by force on this planet, in more than one sense—over-emphasis on Newton's laws. Force is not basic.

Regarding Newton's laws and the mechanics of moving bodies, as indicated earlier, just as we know in quantum experiments that until a measurement is made on the momentum of a micro-particle it cannot be said to possess momentum, this applies to second-order coherence in which macro-bodies do not possess momentum mv (mass times velocity), prior to measurement. This is due to the fact, as mentioned previously, inertia is contextual, and therefore also is mass. When we have the context of a surface to surface (2D-3D) interaction the latter will collapse the one-body coherent state with space-time, causing the properties of force, inertia and mass to come into existence. This clearly also applies to kinetic energy ($\frac{1}{2}$ mv^2), inertia and force.

A body in free fall is coherent with space-time until acted upon, destroying the resonance between the particles of space-time and the body's structure. Further, the gravitational field is not a Newtonian force field (which acts on the surface of bodies), it interpenetrates the body (as per scalar fields) down to the nuclei and the body simply goes with the flow until interfered with.

Let's take other so-called force fields. The 'electric field' surrounding a charged mass is not a force field until a test charge, say, an electron, is held there, which then reveals a force acting on the test charge—the force field didn't already exist (contrary to science). Release the test charge from any constraints and it immediately becomes coherent with the scalar field of the charged mass.

To preserve coherence there are two possibilities:

1) Retain and maintain any naturally existing quantum coherent condition. This is what science is endeavouring to do to enable the quantum-computer condition to be attained.

2) Achieve quantum regeneration using scalar-field technology (4D-3D interaction) to create, ensure and hold a quantum state of coherence, such as in the previously described propulsion system. Also recall the very different example of mind coherent states (mental/emotional/aesthetic) in art appreciation in which the viewer quantum regenerates with the mind, a quantum state (of unity, coherence), and holds it during appreciation—which is instantly decohered when switching to the intellectual observation. Another very familiar and practical example is the laser. But science does not recognise that by putting the noncoherent rays of light into coherence it actually quantum regenerates a single new coherent state of higher frequency and of which the detectable rays of the laser beam are now subharmonics.

Notes

1. www.nhbeyondduality.org.uk. Fractal Matrix articles.

2. William Brown, Nexus Magazine. Article: *Morphic Resonance and Quantum Biology.*

3. Book: *The Attainment of Superior Physical Abilities.* Noel Huntley.

4. This was briefly explained originally in the author's experimental psychology doctorate: *The Mechanics of Learning: A Holographic, Nonlinear Theory* (Kensington University).

5. Ibid.

6. Book: *The Attainment of Superior Physical Abilities*, Appendix C: Mathematical Analysis of Physical Mobility.

7. www.nhbeyondduality.org.uk. Article: *Action Concepts*.

8. Ibid. Article: *Holistic systems*.

9. Ibid. Article: *The Basic Energy Unit: The Vortex*.

10. Similar holographic theories are being researched today under Holographic Principle and Quantum Holographic Storage; see Internet.

PART EIGHT

FURTHER RELATED TOPICS

39.

QUANTUM TELEPORTATION

Quantum Information shows that it is the transmission of information only and contrarily no bodily teleportation takes place. However, the New Science recognises geometric information and the above 'contradiction' is no longer applicable.

The reference to teleportation is an idea that occurs quite naturally from the application of quantum information, and merely serves the purpose of an interesting and entertaining but nevertheless usefully intellectual feature of quantum information. It is a very special case, and is referred to as quantum teleportation (or entanglement-assisted teleportation). The reader, however, if not already familiar with quantum teleportation, will likely be disappointed with the explanation. It does not remotely resemble the science fiction idea or, in fact, the classical physics notion of deconstructing the body to be teleported, and transmitting linearly the particles and information (programs) to be reconstructed at the other end; an incredibly unlikely and dangerous process. Contrary to the familiar but precariously dangerous teleportation, quantum transmission involves no transference of anything material from one location to another. Nevertheless we shall see that our New Science interpretation is closer to quantum teleportation than either science fiction or current scientists' notion of what real teleportation should be.

An electron is a typical example to use for demonstrating quantum teleportation. All electrons are found to be identical (e.g., mass, charge), but, for example, spin characteristics can be different. Therefore if we have two electrons, some distance apart, and we transmit the spin information from the first to the second, this is equivalent to teleporting the electron. The second one is now identical, including spin, to what the first one was, which is now discarded. Thus in a sense scientists are correct: transmitting quantum information can be considered to be teleportation.

Of course, no mass or object has been transmitted, only information. Only the characteristics of the object have been 'teleported', though with the difficulty of needing another corresponding object to transfer the properties into. Also it is difficult to envision this being achieved for complex objects. Nevertheless the surprise is that transfer of quantum information is more correct than we may initially think. However, compared with the real situation, in that we expect mass to be transmitted, it would appear to be an inadequate effort. Nevertheless, not only is it genuinely transmitting information, as mentioned, it is closer to true teleportation (or more common universal principles) than would at first be suspected. But we are back to the same old problem; current science explores the universe as linear, and linearity takes us further away from truth; reality is extremely nonlinear (Figure 1).

An interesting clarification arises when we take this quantum information approach to teleportation, which shows how close in fact the idea of quantum transmission is to true teleportation; that is, the idea that only information is transmitted. Surely we are contradicting ourselves here; real teleportation transmits mass from one location to another; that's more or less the definition of teleportation.

Firstly, we can say that if everything is geometry (frequency patterns) and quantifiable, then we can say that every manifest thing (which is made up of particles and waves) is physical (even if it is energy). Note again that our software and information that run our technologies are non-physical. For example, the spin characteristic of the particle is not physical; it is a description. Now imagine a real teleportation experiment. The object consists of a conglomerate of atoms. In general, a macro-object (of multiple atoms) is not totally in a state of coherence; all its atoms are not entrained in frequency, that is, not in phase or resonant. If they were, it would have one major oscillation for the group: a carrier wave (New Science).

The Emerging New Science

Now suppose this state of coherence of all atoms is achieved (a scalar generator with spectral output could do this). This second-order coherent state links with quantum coherence of the quantum realm. One does not have to think in terms of the impractical idea of disassociating the object into atoms or particles, and not only sending these over a distance but to have copied all the information required to reassemble the object at the final location. We now have a coherent collective oscillation of the object and if we increased the frequency we could take the object out of the 3D spectrum. It has moved up the spectrum of frequencies—recall Figure 1. Also this is a similar process to the example given on spacecraft propulsion (Figure 5). Now further recall we mentioned that every particle has its own location coordinates—and the total universe can know instantaneously where any particle is. Thus if the coordinates of the location are programmed (which is geometry) into the coherent waveform of the object it must instantly appear there at that location. This is the nonlocal phenomenon; the matching resonant condition of the two wave patterns demands it.

But this appears nothing like quantum teleportation. Where is the similarity? The similarity is that the object didn't really move anywhere (in both cases). It blinked off, then the complex coherent waveform was reshaped and its carrier-wave frequency reduced, which corresponded to the new location. So it was information after all, which 'moved' it. But the information is in the form of geometry. The geometry is physical but not the information that transferred it, which was merely resonance of coordinates—and we have the quantum phenomenon of non-locality or phase-entanglement.

Remember the complex wave gives the relative location of its parts or particles within the span of the whole carrier wave (Figure 12). The New Science in fact indicates that all motion is only apparent motion. In general, particles or atoms in motion (even when not coherent as they were in our teleportation example above), as they oscillate on and off, reappear in a new spatial position—similar to moving neon-sign lights, which are simply lights blinking off and on in sequence, giving the illusion of the lights moving.

Nothing was transmitted linearly across 3D space. The object, if of artificial construction, does not normally have a whole coherent quantum state and cannot be 'removed' from 3D and retain its integrity and control. Its frequencies must be raised to take it above the 3D spectrum (and it becomes invisible). To do this, the total body must be made coherent—a single state or frequency pattern (which can be complex). In effect the 'shift' occurs in inner space (with different space-time conditions—even no space locality or time at the most basic level). The body is quantum regenerated to a higher level, the coordinates are matched to the new location, and then the body is quantum reduced, to appear in that location.

We can see similarities to the quantum teleportation comparison in transmitting mere information and no bodily transmission. Thus although certain features of quantum teleportation are progressing towards our New Science, interpretations are becoming more cleverly quantitative and taking us in the other direction. Science's confusions arise from this very mechanism. A purely quantitative linear procedure itself will give rise to the paradoxes and infinite regressions—science assuming that all observers are a mechanism of a similar order to that which is being investigated. The experiential and geometrical approach resolves this since the universe's programs don't function like our computer software. Quantum computing is closer to the universe system, but is limited by the investigator's experimental instrumentation.

An interesting further confusion arises in this discussion, that in emphasising information and software, which is what we might call an intellectual system, in a manner, supports a hypothesis of an intelligent creator (but of the 'human' thinking type), the exact opposite of what science would hope to be the case.* We have alluded to this before. Our computers operate on algebra: a representational, relative system (let A equal this or that). Or, for example, the word 'table' is not a table but represents it, and it requires the human mind and intellect to translate it. Anything representational entails a translator, thus supporting (unintentionally) an intelligence behind the scenes. But at the same time the scientist knows that intellectual representation, algebra, instructions that require translating, are not

The Emerging New Science

part of an absolute system. The approach must be fundamental, no preferences or biases. And ironically the system we are presenting, the geometric mode of intelligence, *is* absolute (intelligence communicated directly by patterns) and would befit science's view of seeking no creator much better. These arguments are called tests of truth in physics (such as generalisation of a theory or symmetry). This is good physics. However, we still hold to the requirement that the geometric intelligence, which involves direct perception, must still be given initial guidelines for any evolution to go towards greater order.

Thus the above problems of both quantum teleportation and imaginary real teleportation take on a new light when we recognise that the universe functions on geometric intelligence. Is all existence geometry, that is, hardware? This at least adds some validity to the quantum case (and preferable to scientists). We are implying that in 'real' teleportation the object is 're-moulded' elsewhere. It is as though in real teleportation the object didn't actually move but the oscillations of the parts blink on in a new unit of space. The analogy of the ocean and waves is still a good one, in the sense that a wave does not bodily move along the surface but the disturbance is recreated continuously. This is somewhat similar to the scientist's transfer of information only, but in our geometric case we do not leave one body behind that has to be cancelled out when the one at the receiving end is recreated.

Science is not taking the subject of information far enough. The scientist's quantum information explains transfer of all traits from a body to another body and this equals quantum teleportation. The New Science recognises that the atoms and the body itself are information, it doesn't carry non-physical information like our computers. Now since all particles blink on and off, we have the potential for genuine teleportation.

* If we have a software method, someone, some intelligence must have decided on this particular encoding system—that's what we mean by arbitrary, relative, preferential. These are all unacceptable physics arguments. But, on the other hand, we are not saying our geometrical intelligence has not been guided; just that it would be easier ironically for atheists to use the absolute characteristics of geometric intelligence to remove a Creator than use software systems.

40.

FREE WILL

The deeper one looks into the subject of free will, the more it escapes one.

Initially, the meaning of free will can be put in the simple form that if one can choose and the choice is not predetermined in advance, say, by fate, or God, then one has free will. In analysing free will, we are not interested in any type of mental blockage to free will, such as from brain-washing, dogma, or programming. Even if the mind was totally moulded to obey instruction/programming we are not interested in the lack of free will caused by this (it is not addressing intrinsic free will).

Let us first examine the scientific standpoint on this. Prior to the development of quantum physics the reigning theories were predeterminism and Newtonian physics. All future events were considered deterministic if there was full knowledge of the initial conditions, no matter how complex; even human behaviour. However, quantum theory was established and is now recognised as the most advanced science. Observations at quantum scales reveal quantum uncertainty, which is fundamental no matter how much knowledge we possess of the initial conditions. But taking a larger scale, the macro-level of the environment, we statistically again have deterministic events; that is, quantum theory reduces to the classical result. At this point, of course, science concluded there was no free will. However, how do we know the mind and consciousness follow the same rules? Science places both mind and consciousness as effects and not causes and thus justifies ignoring, in particular, consciousness as we have covered previously.

Let us continue with the scientific view. In considering the validity of free will, it clearly involves consciously making a choice from known items. But we might choose one item because an existing mental/memory pattern may give a feeling of attraction for it. However, if we can still deliberately think of changing our mind and choosing another, then this looks like free will. Nevertheless, we may say that there is a program to do this, even to change our mind deliberately. This could be a program built up from experience that biases the choice, good or bad, all controlled by the programme—the stored data of experience. The program may be a generalised one giving preferences on the basis of survival for each individual, or even for the greatest good (creating ultimately a good person).

Thus we can always say there is a prior determining factor and we finish up with a deterministic result with no actual free will, merely one which is potentially predictable. Alternatively, we could take the view that all is random and we merely experience the illusions of choice, where consciousness appears as a consequence of the random choice, rather than as a cause. However, if it is random it would simply be a series of reactions, one following after another by chance (and consciousness simply manifesting as an effect). The 'random' interpretation clearly does not align with human experience in which thought and action is mainly organised (scientists appreciate this), which means the randomness viewpoint would have to rely heavily on the existing material laws of science to explain any organisation and human behaviour. However, when we look at the problem of free will from purely the scientific quantitative view (as scientists recognise), there can be no free will in randomness itself, and none in determinism. But a combination of randomness and determinism would be nearer the truth.

Quantum information arrives at the tentative conclusion that free will similarly is neither deterministic (which means: predictable, fated), nor random, but somewhere between the two and that free will cannot be proved. This may appear like no free will to both sceptics of free will and believers, since consciousness is not playing a causative role. However, we can show that there is a practical solution to this if we take a big enough picture. We are back again to the same old theme—and using the same analogies—that we will never understand the complexities of life and its causes by viewing only the twig level (to understand the tree) or only the ground floor of the company organisation (to understand the company). More of this later.

Thus quantitatively, free will would appear to be between randomness and determinism. However, this is all based on a linear 'one dimensional' universe that is formed merely of particles stuck together by forces—Figure 1(a). The subject of free will has hardly begun.

Further developments in quantum information have revealed that in considering the random/deterministic approach, they could be combined. When we combine many random things, the outcome can be predictable in special cases, both classically and quantum-wise. When tossing a coin, whether heads or tails comes up, it is random, but statistically with large numbers, say, one hundred throws, we can expect a fifty/fifty distribution of heads and tails, and we have determinism.[1] In the small-scale quantum special case, experimenters have shown that two or more random set-ups (with probable solutions only) can be combined to create a deterministic result—a certainty. But again science is relentlessly endeavouring to replace consciousness by showing order can arise from randomness, resulting in large-scale events to explain existence.

Thus the quantitative approach of science is to focus only on the parts in an assumed objective world in spite of what quantum theory has revealed, regarding all observations being contextual. The observer is continually being ignored (but in contradiction, is unconsciously assumed, which irrationally makes the quantitative approach 'feel' plausible). Furthermore, how can life or consciousness ever be explained by leaving out consciousness. Can we explain the student's progress by leaving out the teacher. Imagine a car with driver in a wide open space with terrain and environmental variations, such as wind. Can we explain the behaviour of the car by leaving out the driver? It will never prove the presence of the driver since it was the whole point to leave it out. Some explanation would no doubt be found for the car's behaviour even with the omission of the driver. It is thus necessary to analyse the observing process, and we are back to the swivel chair analogy.

There can be no existence as we know it using the current scientific theories and models, that is, the quantitative, linear, 'all is particles stuck together' approach. On this basis, science can only elicit discussions of randomness and determinism, which only produce a string of automatic causes and effects. It can't be anything but a lifeless model, intrinsically, without the full nonlinearity of superimposition (potentially infinite). But let's be a little more fair, and acknowledge that science is not claiming to do any more than the above and that it considers consciousness, life and awareness, to be illusions anyway. The catch though (mistake being made) is that science is using an observational process to find out these things that fully includes consciousness—a complete contradiction. The consequences are to cause science and intellectual observations to quantum reduce (collapse the wave) that which consciousness is really observing, to the commensurate quantitative level, subsequently proving that science is therefore correct.

Nevertheless in support of science, every decision, or choice could be biased (caused by prior information), which contradicts the notion of free will. And further, one could say 'No' to the choice, but there may be a program for this also—how did it get there? We have to keep placing a cause behind a cause, and we have the infinite regression—totally unacceptable to science (as it should be). However, science places the regressing causes on one line—in the same dimension. There is a fractal system of dimensions here to utilise for our explanations that goes inwards—within.

With the new system, ultimately the regression now goes back to the whole or source, which could be infinite and the riddle then doesn't exist. We don't have to have a dominance of rigidity in the part (a predictability that clearly lacks free will)—fractal levels give randomness, freedom and determinism. The part (for example, the individual) is still choosing within the fractal restrictions, which are greater in the lower dimensions (a ground-floor worker has less freedom than the manager, regarding company knowledge and operations). We know the structural mechanism of the fractal; for example, a branch has

The Emerging New Science

greater wholeness than a twig. We are indicating here that even free will is fractalised (corresponding to consciousness and its higher aspects). We have to consider endless possibilities of choice—the possibilities increasing into higher fractals to our previously considered absolute of infinite possibilities (higher quantum realm).

We have considered free will based on randomness combined with determinism. This becomes much more complex with superimposed levels of randomness/determinism as we include the hierarchy. Now if there are endless possibilities, choices within choices, we can call it 'free will'. In any case the creation point could have very broad guidelines to follow; for example, a blueprint to have the drive to exist (in a particular form), like a mould, an experiential template. An experiential template would contain broad or general programmes, allowing choice, within it, even if these choices are then based on sub-level programmes, which can be a complex network of preferences.

In this argument on free will we could always simply bring in that magical beginning, God, to explain everything and there is nothing else to puzzle over, but this is not science, and at our level of awareness it is essentially guess work (it is one thing recognising an initial intelligent source and another that it runs everything, including free will). There is a direction of thought that states free will has been given to humans by Source.

The fractal system, however, controls the degree of freedom and will give us less degrees of freedom or choices on the lower levels. The next fractal level up will have more degrees of freedom (less restrictions from fundamental templates (imprints/programming) and so on back to the infinite degrees of freedom, at the absolute).

Now we can use the ocean example. At the absolute there are no imprints, no wave patterns, vortices, all is calm, still. But the infinite possibilities are still there. It would initially appear that anything could be selected, with no predetermined stimulus, such as a preference. If there is a preference then it would mean on this basis there are more hidden causes. We have discussed in the earlier section that a portion can be separated off, creating objectivity and an individual viewpoint. However, for a start there is a natural preference even at the absolute. Since the infinite absolute is one undivided whole it is the ultimate essence of harmony (love); separation does not exist. It could not think negatively (until these experiences had been built up in and by the lower levels).

What we are doing here is we are conforming to scientific procedures (not necessarily beliefs though) to see if we can arrive at a satisfactory explanation of free will. Scientifically, as we continue with this approach it can be considered quite mechanistic and within the quantum realm of infinite possibilities. Using also the description elements of the new approach, we can now consider that this absolute creator from a totally subjective state creates not just templates for matter (already mentioned earlier) but experiential templates that contain multiple possibilities, which directs the lower conscious levels. A primary template program may be, say, to behave and act for the greatest good. Another primary program could provide a drive to simply exist—in a particular body form and dimension. At the mind level, randomness provides the material to be moulded into experiential templates.

What we must arrive at is the solution that all decisions and apparent free will choices are made inside the context of the whole—which is basically known by all. The part, potentially, has all information of the whole, but makes choices for itself, the part—no paradox. The part is referencing the whole. There can be no authentic choice (selection) unless the elements of choice are spanned simultaneously—recall the swivel chair analogy. The selection made will then be based on the parts inner data (experiential templates, programming).

We could describe free will as the ability to move one's attention randomly over many items (in the mind, say) and then select (even though the selection was determined by stored experiential data). The infinitely nonlinear state gives the appearance of free will. The swivel chair analogy is relevant here, that is, we must have two viewpoints (or more) simultaneously in order to choose one. With nonlinearity there is a huge amount of data from local to global in a fractal system of contexts for selections.

How does selection take place? Sounds obvious, but it isn't. Recall that the difficulty with logic is often because it is too 'simple'—is it assumed—the assumptions are ingrained and part of a fundamental context that everyone develops and becomes unconscious of. We will use the swivel chair analogy again and this will also show that larger quantum states are primary and smaller ones within the larger are secondary.

To recap, we have the thought experiment in which we have a person seated on a swivel chair. We imagine his mind has been cleared of all recordings and memories, and there is only raw, unmodulated consciousness—sheer awareness. He is in a fixed position looking ahead at object A, and he can only see the object. This is his existence, his universe. As though in a hypnotic trance, he could be stuck there indefinitely. Another individual now swivels the seat rapidly by 90 degrees so that the subject is now staring at another item, object B, and nothing else. Since no recordings are made, he is simply using a blank state of consciousness. For the scientist, this is a quantum state of energy, but we can allow that his focus could reduce into the details of the object.

The key item now is that in order for him to be able to be aware of both objects A and B, he must not only have to record A and B (separate occasions) but must also record A and B, not merely stuck together (associated), but as a new state C; where C is not just the sum of A and B but a new state C with A and B merged together as one whole. Note the resemblance to superposition in quantum theory—though this would be better described as superimposed. While the subject is thinking of A he can also hold in his mind, C (because A is in C), to enable him to be aware of B and thus be able to switch his viewpoint to B by going to C first (assuming he can turn to do this).

The ability to choose A or B, or just think of B while looking at A, without getting stuck on one or the other, depends on viewpoint C, the simultaneous quantum state of A and B. Clearly C is a superior state to A and B—it is a higher order. Even if C already existed before A and B it could be quantum reduced to A and B. The ability of consciousness (these quantum states of energy) increases up the fractal system. The larger the span of this state, the more advanced or evolved it is, that is, the greater the whole (hence 'holy' in religion).

All we need now is sufficient complexity of quantum states and their associated frequency patterns to give a satisfactory rendering of free will. It does not matter if it doesn't strictly qualify as free will. The complexity is many layered and goes back to the absolute and the whole. At the absolute we postulate that there is an experiential state that knows 'everything', totally governed by feeling—no location, not quantitative. Like a single state of knowing everything at once (but not intellectually). It is one-hundred percent subjective but with increasing degrees of objectivity down the fractal systems of increasing separation of parts (portions of consciousness). See Appendix C. When the whole exists in the background, this is a potential condition for all parts, individuals, on the lower levels to reference it knowingly or unconsciously. Science only takes the lowest fractal level.

Thus the fractal system not only handles the infinite regression but also free will, as well as possible. The fractal levels are the deterministic factors and the 'freedom' between the fixed levels is random but available for organisation.

In terms of a reconciliation of the orthodox view and the New Science, the energy states of observation can be called consciousness. But it will now also mean that in some way all energy is consciousness—even an atom has a minute degree of consciousness and intelligence.

The key is the state of knowing all at once and that this is experiential—pure sentience can choose and put on limitations of objectivity. If science says there is no free will, since each choice is itself caused, it assumes unconsciousness is dominant. We must go to the absolute level of total awareness where all is experiential/subjective at once. Think of choosing, then we have what science is saying (that something else is influencing the choice). Then this influence is unknown and must be put into known context all the way to the top (the all). Again the swivel chair example is needed to clarify this. Science is constantly assuming there is a background state of thought (recall $C = A+B$), but not acknowledging it and in fact denies it when we apply it to consciousness. The quantum states (perception of a reality), A and B, cannot connect without C. Without C and its endless (pyramidal) multiplications science can only be quantitative, and by this we mean everything is a mechanistic string of events, running on the automatic.

In the swivel-chair example, the larger quantum state of perception, C, is bound to be more powerful than the parts A or B or, of course, A + B. In psychology protocol the meaning of A + B is simply some kind of association between them. These are merely linear attachments that science is dealing with; they are meaningless, unless they (A and B) have a program to relate them. But we are back to mechanistic chains of instructions devoid of the experiential factor, which we are endeavouring to show is in the whole perception states. Thus we need C. A good example we have given before is the well-known Pavlovian stimulus-response exhibited by dogs, simply referred to as an association between the stimulus, the sound of the bell, and the response, the saliva. The background quantum state C has been omitted in science. This very principle is what great art and music is all about. But of course as many people know we are losing sight of what art actually is.

The infinite absolute is within each single part. Any logic that questions free will—and this would be expected—must, however, be expanded through the fractal hierarchy to the limit or absolute or whole.

Consciousness, including all its higher aspects will contain virtually infinite contexts to cover every conceivable state, either partly through enormous experience of consciousness' various parts (recorded in a 'spiritual genetic pool') or/and simply the infinite possibilities of the quantum realm.

The essence of free will can only be, in the limit, the relationship between the part and the whole in a completely nonlinear manner, that is, the part is selecting/choosing from the whole of effectively infinite possibilities. On lower levels of evolution, experiential templates are more restrictive. Thus at this level, all actions and thinking could be the resultant of the mass of imprints 3D consciousness has received. However, superimposed on this is the influence of the next higher-dimensional fractal level. There will be more influences from unconscious data (giving rise to biased and 'controlled' selection—see later analogies for this). This 'infinite' regression to higher levels though stops at the whole.

The mind sections on these fractal levels will contain an enormous complexity of context—masses of information automatically referenced by the individual. The information is intellectual, detached memories and learning imprints, and also countless emotional states—for example, likes and dislikes. Many of these states coalesce into a whole quantum state of energy or frequency pattern—giving a background feeling about things. They can be quite unconscious but form a virtually infinite (if we include higher levels to the absolute) sea (of consciousness/mind) for impulses to spring from, or decision-making material.

Essentially what we have done here is to take science's analytical procedures for describing free will, but increase the complexity potentially infinitely. This creates a wide variety of different individuals each with a virtually infinite sea of data to compute on.

Let us now consider the analogies we alluded to earlier for demonstrating how one fractal level or degree of freedom is in the context of the next higher one. Thus we are now considering relative free will, such as that in our 3D existence, even though ultimately free will is much greater, given by the fractal dimensional system.

The first analogy is a simple two-state one. Imagine taking a dog for a walk. The leash, linking together the human and the dog, is quite long. The human decides where they go, that is, this higher order level (of control) of the human determines the overall path, the direction the dog takes, but the dog has choice and free will within this restriction or context. The dog can even get into mischief; it has freedom to move around, governed by the length of the leash (we should not disallow the human to additionally pull on the leash if need be, even within the leash-length limitation). Thus the dog can represent the human lower consciousness; an extension from the higher, or soul consciousness represented by the human with the leash. (Keep in mind this higher level operates within still higher levels but with lesser restrictions.) We see that the human's free will is superimposed on the dog's free will, and can, broadly speaking override it. There is a nonlinear relationship between these two variables.

We might also wish to note that the lower human consciousness represented by the dog could be programmed to not recognise that there was any limitation governed by the leash. This means the perception beyond the leash length would also be out of bounds—nothing would be detectable beyond this limitation with this supposed programming. Thus the 'dog' would think, in fact, it has complete free will—an illusion many people have today. We see from our standpoint that freedom is within a set framework. This framework can be considered to be a fractal level of consciousness. There is free will relative to a given framework, such as this 3D reality.

The fractal levels provide the boundary conditions for free will—a finger relative to its own joints can't move as freely as, say, the shoulder joint would allow the finger to change position. In effect, fractals form closed systems of standing-wave structures.

This would include fixation of the mind through programming and brainwashing. Free will is always relative to a framework or context. Maximum free will would exist at the greatest whole, God, or the Infinite. This level would provide maximum probabilities for selection at the lower (fractal) levels. Note that these lower levels, although limited in the above-described manner, could select probabilities outside their fractal-level boundaries, providing the subsequent ramifications of this selection is in full alignment with the corresponding higher purpose or greatest good—the dog is, say, allowed to (choose to) go over the fence, providing this is within the possible purposes (paths) of the human.

The ocean wave analogy is appropriate when we take the quantum information free will point of view. We have electromagnetic waves propagating through space, say, radio waves, modulated to carry information. Thus waves are real but the modulations are information. Whereas we, with our information systems, computers, algebraic coding systems, would designate a modulation to convey a meaning (quite artificially); in contrast the universe would utilise the geometry of the wave. Both types are information though. The substance of the waves is still there even when the information is erased. Compare ocean wave. To restate this, the ocean is real, substantial, but the modulations are information—in this case the geometry counts as information. But we can now see that all could be explained in terms of information, as indicated by quantum information.

Science, in stating, if there is order in the universe, follows with the question, is there a need for free will? But science is assuming its laws apply to behaviour of life. Scientific laws apply to the immediately detectable environment—the third dimension—whereas the mind/consciousness extends or has some link with its higher aspects involving higher laws of physics, much greater degrees of freedom (more paths) than, say, for instance, the rigidity of a planet orbiting a star (the physics law of this).

Furthermore, scientists can never make any sense out of these things until whole states (of information) are recognised. Such as is indicated by quantum theory, in the form of quantum states (unity), superposition and superimposition, and the fractal holographic system.

In the swivel chair analogy, science in ignoring consciousness and the full implications of the principle of context, which means in this analogy, it is omitting the perception-quantum state 'C' (A + B) and can only arrive at randomness and/or determinism (note the triad principle referred to earlier). It conceptually quantum reduces the higher states of thinking required by the imagination and intuition (properties of consciousness) and destroys what it is searching for. Recall earlier examples in discussing collapse of the wave, such as in art and intellect.

Thus as stated, science is always correct with respect to its chosen context. But as one may recall from the earlier section, anything will self-prove relative to its context.

We commenced with an opposite viewpoint to science, in particular, that consciousness is primary, and we have developed at least a tentative conclusion by extending the initial valid arguments of science through the fractal hierarchy up to the absolute of infinite possibilities. In other words, by taking the scientific viewpoint on this to the extreme we tend to do a full circle and essentially resolve the conflict with science. The

The Emerging New Science

second factor that we set out to achieve was to show that larger quantum states control smaller ones and not vice versa—life/consciousness is top down (large to small) and not bottom up. Those quantum states of greater coherence provide a platform for consciousness, but they can nevertheless be quantum regenerated.

The final conclusion, it would seem, is the following: either 1) we have to admit that free will goes into the category of nonquantifiables, such as the experiential, sentience, and the 'aliveness' characteristic (which can't be understood by finite, quantitative, methods); or 2) we apply the infinitely nonlinear quantitative analysis, utilising fully the holographic and fractal principles. To make one behaviour different from another, we might have an initial all-encompassing programme, slightly different for different subjects ('offspring'), such as one given the perfect holographic principles, which means the subject will always act for the greatest good (whole within part feature); another has a slight discrepancy in the holographic configuration, enabling them to take more devious and less positive paths, and to increase variety we can combine this with a background fractal randomness in the lower levels. Thus the complexity (infinite nonlinearity) itself will give the appearance of free will.

41.

GEOMETRIC INTELLIGENCE

Geometry is the universal communication system and brings thinking, perception and understanding into the direct and experiential focus of consciousness, free from representative systems, such as most intellectual activities associated with the educational system.

We have made numerous references to the universe as functioning on geometric intelligence, which would be expected to include mind and nature. All these states, mind, nature, universe, would embrace consciousness, in particular, since we are recognising consciousness as primary. However, we are not concerned specifically with mathematics, arithmetic, or algebraic geometry, but mainly a discussion of the philosophy and psychology of consciousness' relationship with these properties of life and existence. On this planet, civilisation is mainly familiar with intellectual intelligence, which is based within the principle of objectivity. But objectivity, which is the same as degree of separateness, is relative and is thus on a scale with subjectivity. We have seen that subjectivity can be 100%, and must be, at the Absolute or Source. This condition is the origin of geometric intelligence and is more difficult to understand than the familiar objective intelligence that manipulates labels, symbols, language, equations, graphs, and creates the coding system used in our artificial computer, which is non-physical software and, as we have stated repeatedly, does not depend on geometry but follows algebra-type principles; these are representational.

We nevertheless use, to some degree, the geometric mode with right-brain consciousness perceptions, imagination, intuition, but it is usually merged into the intellectual ability. Thus we shall describe how the geometric system arises by beginning with 100% subjectivity (the absolute), which is not entangled with the intellectual mode.

For the sake of convenience, let us take an imaginary portion of the infinite condition of the absolute with the subjectivity at 100%, meaning no parts, no separation, divisions, and there is no objectivity. Its beginning state is one of potential with the infinite possibilities (a parallel with quantum physics). Recall Section 25, Figure 11. This scenario is easier to picture than the infinite but we must realise anything it creates such as a particle, must be inside itself (there is no outside). Thus in order for one part of this selected portion to perceive a particle (another form of itself) it must temporarily block its view from the particle, and only view from the chosen viewpoint. Immediately a degree of objectivity has been introduced (separateness). Imagine building up complex configurations in this manner—an environment.

The condition of the total portion might now be 75% subjective and 25% objective. Note that in our civilisation it might be something like 99.5% objective and 0.5 subjective (viewed and experienced by the 3D linear conscious mind). This high level of 75% subjective for our selected portion is sufficiently high that it has not forgotten that its total self includes the environment.

We can now see that the viewpoint can (by recapping on its knowingness) choose to be all that is objective, separated-off elements forming the environment. No 'thinking' or mental activity is required, no intellect, only duplication, involving shapes, patterns of energies; hence the use of the term 'geometric intelligence'. Thus it can simply know everything by being it; since it is all consciousness.

As a thought experiment we might imagine increasing the coefficient of objectivity (ratio of objectivity to subjectivity) by increasing the degree of the separation factor in a particular existence. This would mean the average magnitude (size and frequency) of the 'building brick', the quantum (coherent) state, is reduced. There will be found to be fewer regions of coherent states, both in the environment and the mind and consciousness.

But this is not the main reason why science rarely encounters them, as indicated previously, the scientific approach quantum reduces these coherent states to their parts, giving a lower level of truth.

As an additional feature to presenting evidence for the geometric approach we can include the ocean analogy, given earlier. The ocean represents the absolute, infinite, stillness or Source, and the wave/vortex patterns on the ocean are the manifestation of all quantifiable phenomena, matter, thought, emotions, mind and consciousness, where these are expressed in particles and waves. In this analogy we can now 'see' the modulations of the ocean forming the wave-pattern structures, the basis of an existence (but we can't see and science can't detect the 'ocean'. The modulations are shapes, something moulded, and we have information and intelligence from geometry.

Scientific theories are becoming lost in mathematical abstractions, non-physical software, or information (quantum) theories, intellectual and logical representations, and we are back to number theory versus geometry, in which we are stating that the universe functions on geometrical intelligence. As repeated many times, this quantitative approach will draw science away from recognising the primary causes in the nature of experience, sentience, and consciousness. A confusing aspect here is that primary causes are non-physical, but not in the sense of non-physical software patterns of, for example, our computers that merely have mathematical significance. As we have seen from the ocean analogy the primary condition, the infinite potentia (compare quantum realm) is very 'substantial'. However, clearly numbers play an important role;[2] if a universe is organised it must have mathematical consistency. This is clear in geometry. The geometry of the universe/life is not primarily random and therefore numbers arise in the evaluation of exact proportions. Some proportions are not in line with this cosmic creation and are what we would refer to as nonharmonic, and also finite, not self-perpetuating through a continuous source.

The intellectual mind, as it stands today, has a predilection for abstractions, representations (algebra) rather than dimensions and structure and geometrical bases for life and the universe. Our subconscious can understand symbols, which may

The Emerging New Science

have geometric significance. It is much more graphic than the conscious mind. For example, a negation is abstract—the subconscious does not recognise this and will only focus on the picture presented. To tell the subconscious we don't (a negation) want something, simply causes it to focus on that 'something'.

The conscious mind can't understand directly the geometric significance of the universe's structures (such as blueprints, codes). But our cells/DNA (which are, in fact, part of the subconscious) can understand this with the occurrence of resonance. The mind basically processes pictures and patterns. It may come as a shock to people to know that the subconscious cellular information can read the crop circles. All geometry radiates wave patterns. Crop circles created by benevolent visitors radiate harmonic wave patterns containing geometric information that resonate with the DNA to aid promotion of a more optimum evolution. These were placed mainly to compensate for other circles that created a harmful geometry. Human crop-circle creations would of course be random and therefore garbled in meaning. Creation itself is thus based on shape of energy; this is information. It involves direct perception and therefore direct experience (thus 'moulding the mind' is more literal than one might suppose). Potentially, this is the ultimate way of knowing; by experience, by actually being the thing to be understood. Compare our intellects, which function objectively (indirectly) and thus use representation, such as languages, algebra, and manipulate abstract information. Geometry is a universal and absolute system. It would be unthinkable that any other, or relative system, would be used by the universe or Source, and the sphere has the fundamental geometry; the basis configuration of creation is spheres within spheres (referred to earlier).

Thus the subconscious cannot handle the 'not' when given instructions, it would simply put a counteraction there which would hold it there more solidly. To cancel something it must first make sure that something is fully re-experienced, then focus on anything desired/positive. Geometrical intelligence is fundamental and a property of creation; how creation occurs —hence holographic and fractal patterns. This means that in the process of creating a new idea or a new thought, which is substantial and is energy, the thought, say, a negative or undesirable one, can't be erased by counteraction, thinking 'no' or 'not'. This 'error' is the human mode of creating representation, which is indirect, including intellectual thinking. We name things, such as an object, by a label. It is represented by a name, which has no intrinsic association with the object. We don't transmit the energy pattern of the object to another person (basis of a proper language). Thus what one talks about needs to be translated by another. Our computer systems operate on the same principle, which specifically is algebra (not geometry). A pattern of 'bits' (electrical pulses) represents something. To erase something by the mind it must be resonated with, which means duplicating it with one's consciousness and replacing it with the desired aspect. Anything else, any other mental state, will simply add to it, especially a negative, creating a counteraction and lock it into place.

To give a human-level example, psychoanalysts had a certain notoriety for telling the patient the 'cause' of their malady. This sets the mind of the patient to receive an intellectual representation of the problem on which he or she now focuses. This puts the attention of the patient into an external mode, reducing their chance of finding the region in the mind (geometrically) and duplicating it, to solve it. In other words, any cognition, which must be subjective and not objective, is prevented. It can't be solved intellectually. The patient's problem is not a puzzle to be solved. It must be (re) experienced by consciousness. It is a structural pattern that is causing the block. A similar situation holds for religions that teach, even if unintentionally and innocently, that there is an external God, rather than internal and in a path inwards, the true 'within' (see fractal-tree analogy).

All this indicates that there isn't true understanding externally, it can ultimately only be by consciousness 'being' the thing, experiencing it; an ability that has declined on the planet. This is the subjective state rather than the objective. This is telling us something about the different nature of fundamental creation itself, that experiential non-quantifiable state, where everything can be known. Even the 'small' part, such as a human, is an expression of this source, of which the latter is within everyone and everything. Even the individual within a collective (assuming a harmonic condition) is holographically related, and therefore each individual is like a geometric element within the whole—a jigsaw puzzle is another analogy. In fact, there is really only one self (the Absolute/God/Infinite) expressing through different forms. There is no true external state; only limitations imposed to provide objectivity, separateness, agreeing to not know at some level (see Figure 11). This is a counteraction so that one part (of the whole) does not know another part, which causes separateness and objectivity and the formation of regions of unconsciousness (from consciousness).

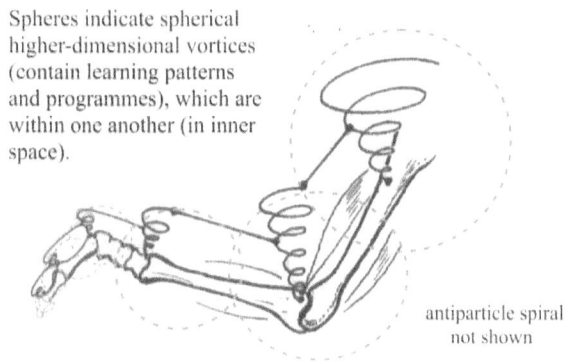

FIGURE 23: Rough depiction of arm vortices. Compare Figure 4.

Returning to the learning pattern, note that when controlling movements, such as in skills, we can simulate this control by considering equally the control over the distribution of a number of particles. These can even be thought of as 'bits' in a computer programme. If one can control a geometric pattern of points (particles, bits), this can be translated, with a suitable mechanism, into movements, such as for instance the complex movements of hands and fingers at a keyboard. One can see, in particular, from Figure 23, that the learning pattern functions on geometry; it is holographic. The vortex spheres are inside one another within inner space (like the tree)—smaller within larger. Recall Figure 22 and the brief reference to the positioning of spheres inside larger ones that can be precisely known from the sine-wave patterns on the surface (we won't go into details on this since it can only be speculation, but also logical and intuitive). Every part must reference the whole by being inside the whole, which is encompassing all information to be located. Compare this with our computers, which would merely give a string of instructions; one nudge and the reference has gone. The robot arm in a car factory, which swings over a vehicle on a conveyer belt that does a single job and resets, has only one reference point. If one stood behind it and swivelled the arm installation by a few degrees, it would of course miss its target.

The holographic learning pattern has a virtually infinite reference system. Recall the perception states A, B, and C in the swivel chair analogy. In order to control the change from A to B, or back, both these must be within C. Quantum states in perception and consciousness require this. It is a natural consequence of this that the mind-computer system, of which its energy patterns duplicate behaviour of consciousness, will use a geometric system for learning-pattern control.

The intellect is like a clever compromise system so that we don't have to keep looking at things, being things with our consciousness, of which the latter would be too advanced in most mental activities (and is primarily the Source's mode of expression). The intellect would think about, say, that black-hole in the cosmos without being there, compared with extending one's consciousness into it, experientially. Intellect is a property of objectivity, separateness, and to have knowledge about things, from a 'distance'. Clearly it could totally go astray and consider that its own representational system of thinking, its theories and specific experimental set-ups that get a result from nature, are truth itself. To the degree that this mental structure does have good correlation, what one arrives at regarding physical perception and laws may become true (unintentionally manipulated but selected by that reality level). When one has a dominance of objectivity one must use the intellect. We build up contexts: huge structuring of a representational reality in the mind. For

The Emerging New Science

example, we build a model of the atom in the mind; we don't look at the actual atom. This type of looking of course is not merely external but would involve a focussing of consciousness into the object and being it, experiencing it.

The female mind uses the experiential system more, in relation to the intellectual, compared with men. Thus there will be more emotional effects due to this direct experience. This is not valued much in a highly objective and intellectual society, but ultimately in higher states of evolution supercedes the intellectual system; the ideal being a balance between them both. The danger with the intellectual mode is that it will think too much instead of looking (in the fullest sense). We mistakenly emphasise left brain and discourage the right brain consciousness characteristics such as imagination and intuition. The latter may relate to psychic abilities, which are more in the geometric category of sensing, feeling patterns of energy. There is direct contact with the energy, rather than giving it names, representations and manipulating labels. A more advanced consciousness could look directly at an atom, actually be it, and thus understand it experientially—there's no guess work. The intellect, as mentioned, must gather external information about the atom and build a model in the mind. A fact not easily recognised or understood is that a true subjective understanding is proof itself. This is simply because, as already described, consciousness is capable of duplicating, actually merging with the object or event in question.

We live in a holographic fractal universe with universes at each fractal level. Increasing fractioning means increasing objectification in a scale going from a theoretical maximum objectivity (the path science takes and thinks it's a true reality) to maximum subjectivity (which will be 100 percent at the original basic state, the absolute and infinite that is totally nonlinear). If communication became more telepathic, recognition of words would gradually be replaced by basic concepts and we have the direct contact again with the energies; experience of the energy patterns. In our society we can't expect to be experts in this mode of communication but unfortunately non-recognition of this by science can bring about an incremental drift towards the quantitative (only), the representational and further objectivity. In the theoretical extreme of objectivity, everything would be robotic, and no true creativity would exist.

Thus right brain relates more to the geometric intelligence, but in our society left brain activity is highly emphasised and relegates the right brain to inferior status (within the methods of the educational system). Of course, we need to use the intellectual method in our civilisation due to the high degree of objectification of the elements of our existence and environment. As implied, our intellectual information can be abstract and have no geometrical shape to its energies, as with for instance our computer programming system. Its information method, using the binary system, does not have shape. In nature, shape of energy is information. Also geometric intelligence correlates with higher ratios of subjectivity to objectivity. We are in a fractal system with increasing degrees of objectivity as we go from higher integration to lower or greater fragmentation (for example, tree trunk to twigs). The greater the separation, the greater the objectivity. In order to form objective existences, the subjective must divide its awareness so that one side doesn't know another side (as described earlier). This also develops unconsciousness. In a typical objective world such as ours, the environment is separate, out there, forming part of our unconsciousness. What we call the intellect must now be developed to manipulate symbols, labels, words, that describe features of this objective, unconscious environment. In contrast, the subjective could simply *be* the objective features and manipulate them—sensing information through geometry.

We develop complex intellectual systems made of 'codes', labels, representations, second-hand copies of things, in particular, of what can't be seen directly (experienced). Thus, activities that are occurring, or are present in the environment/universe, are coded into a mind so that it thinks about these things that it can't see directly. For example, one can think about a pen without actually seeing it or using it by thinking of the data that describes it. A geometric intelligence means communication systems are direct and therefore experiential, but for us with current human ability we would have to greatly develop and expand our consciousness.

Notes

1. Book: *Decoding Reality* by Vlatko Vedral.

2. Book: *Memento—Remembering the Self Through the Geometry of God's Love*, by Milena.

APPENDIX A

THE SOLAR SYSTEM

Originally a twelve-planet system

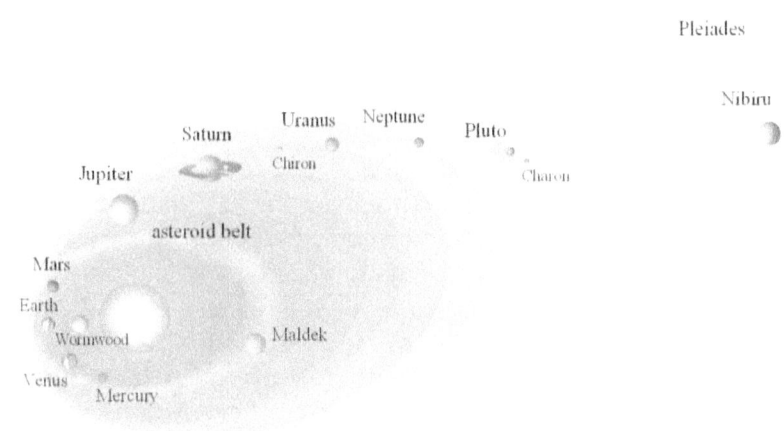

FIGURE 24: Originally were 12 orbiting bodies qualifying as planets: 1) Mercury. 2) Venus. 3) Earth. 4) Mars. 5) Maldek. 6) Jupiter. 7) Saturn. 8) Chiron Charon. 9) Uranus. 10) Neptune. 11) Pluto. 12) Nibiru.

As is known in astronomy our solar system is a lot more complex than most people would realise. Figure 24 only gives the particular features we wish to highlight. There are a surprising number of moons and five confirmed dwarf planets, for example, Eris, which is a little larger than Pluto, both dwarf planets, but there are many more not confirmed. The latter, Pluto, however, is usually regarded as a planet, totalling nine planets in the system, rather than eight. In addition to the Asteroid belt there is a similar Kuiper belt, which is a huge ring of debris encompassing the Sun, mainly ice and contains at least three dwarf planets and a comet in a higher eccentric orbit. In Figure 24 we see that Charon plays a definitive role in the original planetary count. It is Pluto's largest moon, and combined with Pluto the pair is sometimes described as a binary system. Chiron is a minor planet (they total over 500) between Saturn and Uranus, also classed as a comet (of which there are thousands). Wormwood, an artificial planet, is referred to by off-planet visitors as the Nibiruian Battlestar.

We shall only present a brief description of our solar system with the addition of the few very interesting features not covered by current astronomy but taken from what we are referring to as the Guardians' material (transmitted by A. Deane in Freedom Teachings workshops; also in the book, Voyagers, vol. 2).

Modern science has recognised that the solar system is in a chaotic condition, for example, wrong orbits and wrong spins. The precise reason for this is given in the Guardian material, including the Earth's tilt and wobble (see Appendix B), and that the orbital period was originally 360 days (a multiple of 12). There should be 12 bodies qualifying as planets (12 being a harmonic number of the universe construction). A large planet, Maldek, was destroyed (Figure 24), of which the fragments formed the asteroid belt and also other dwarf and minor planetoids, but also, Wormwood, the prophesied body of Armageddon in Revelations. Wormwood, a fragment of Maldek, was artificially made into a planet by Nibiruians from planet Nibiru and as mentioned above is referred to by extraterrestrials as Battlestar Wormwood.[1] This body, sometimes referred to as planet X, was placed in Nibiru's orbit 180 degrees (opposite) orbital position, and used as a relay station. Note that both Wormwood and Nibiru are close to the astral spectrum and they literally dip in and out of visibility with respect to our 3D in their reverse orbits. Their huge elliptical orbits have a period of

3,657.8 years. Nibiru is at present in the vicinity of the Pleiades system.

The original planets, in order from the Sun: Mercury (1), Venus (2), Earth (3), Mars (4), Maldek (5), Jupiter (6), Saturn (7), Chiron plus Charon (8), Uranus (9), Neptune (10), Pluto (11), Nibiru (12). Chiron (classed as a comet) and Charon were apparently originally one body, which was then classed as a planet.

In 2003 Wormwood, in an opposite orbital position to Nibiru, was due to pass (actually was manipulated) between the Sun and Earth, close to Earth, and expected to produce a severe pole shift. It never happened since the Guardians, after only a partial success in destroying Wormwood, resorted to their second plan in which they apparently altered its time line.[2]

Notes

1. Book: *Voyagers*, Volume 2 by A. Deane.

2. www.nhbeyondduality.org.uk. Article: *Experiential Probability Dynamics*.

APPENDIX B

CONFLICTING SPIRALS OF CREATION

A major source of interference within the human race

There are many pitfalls on this planet in one's search for knowledge and this, perhaps surprisingly, is fully applicable to scientific discovery. We have seen that science is limited by the reliance on the experimental method, which can't exceed its own resolution in what it can detect, and further that the observation and measurement process may cause quantum reduction of higher orders, that is, from a state of coherence to a typical 3D random condition, forever proving that material existence is the only one. However, who could imagine that evolution, which in this context preferably requires the term 'creation' or at least 'cosmic evolution', that there is not one source of 'creation' but two rivalling forces. That is, two major currents in opposition, in effect, vying for dominance. At the most basic level these two currents are expressed by conflicting spirals of energy. The natural source is referred to as a Krystal* spiral and the unnatural one, 'Metatronic'. The latter is primarily in a reversed condition (giving rise to reversed polarities) but originated by branching off the natural one, then becoming parasitic of it. This results in closed systems of energy, disconnected from Source. [*We won't dwell on the origin and usage of this word except to indicate that the terms Krystal, Kristiac, Krist are the roots of Christ or Christos. Nevertheless, Christos/Christiac/Christ express the highest dimensions in this time matrix—hence the name 'Jesus Christ' as having incarnated on Earth from this level.] [1]

It is interesting that many religions believe in not only a God but a 'counter' God, a Satan, which in effect threatens to take

The Emerging New Science

over creation. We may see the similarity here, that there is a subversive manipulation occurring at the deepest levels of creation itself; and the truth behind it is quite scientific, to say the least.

Thus there are two basic currents of creation, one endeavouring in effect to harness the other and siphon off its perpetual source, which is the natural, original organic creation based on harmonic blueprints with continuous connectivity to the Source. The origin of this massive interference of natural creation is a vast history of manipulation, involving many intruder races (too big a subject to deal with here).[2]

The subject material of this book, in particular, with a title such as The Emerging New Science, wouldn't be complete without at least a brief mention of the far-reaching effects of these competing 'evolutions'. Such a source of information as this could only originate from a race thousands of years ahead of us—what we have referred to as the Guardians. Further titles associated with it are: Keylontic Science, Freedom Teachings, and Guardian Alliance. The author's book on the original Great Pyramid covers the subject of who the Guardians are and, in particular, the extraterrestrial role in the building of the pyramid.

The Guardian information is telepathically transmitted by Ashayanna Deane. It is not channelled or part of the New-Age movement. A huge body of knowledge has been presented in workshops over the past 12 years. This is just a snippet here but probably the most important at this time on this planet. Note that the main body of this book is not connected to or derived from this source apart from a few brief references, which are acknowledged, in particular, the DNA and chakras.

Unfortunately modern science's standpoint on creation or evolution is one that emphasises closed systems, finite resources and, in particular, a fixed amount of energy for the universe that ultimately dies an entropic death as its potentials reach equilibrium. This false belief system, however, does have some correspondence with one of these creation sources vying for dominance, which would of course be the limited dependent one (the opposition).

The Guardians refer to these two creations as the Krystal spiral and the Metatronic spiral. The growth of the expanding spiral expresses both the cycles and expansions of creation. The Kyrstal spiral has the natural source, which is perpetual and infinite. The artificial Metatronic spiral attaches to the Krystal at some point and acts to pull the natural creation currents away from alignment with Source and follow the Metatronic frequency structures—an artificial growth system with a fractal patterning that follows the Fibonacci series of numbers and finally the golden mean proportions (information which understandably will be severely resisted by golden-mean enthusiasts). The fractal ratios may not be conducive to or resonate with those for continuous creation. The Fibonacci sequence expands by adding the last two numbers and, in effect, feeds off itself; it is not in alignment with the Krystal spiral of creation, which increases geometrically and in which each level is recreated and connects with the Source/Absolute.

Creation has been given certain values, ratios, and harmonics. Any other creation will not be in alignment with its source and will not draw continuous resonant energies. It can't keep regenerating the energy and must be parasitic; feed off itself initially, then to survive, be parasitic of other sources, characteristic in fact of existence on this planet. Much of this planet and its life has been converted to the Metatronic system, which as stated, is finite and parasitic; it can form into black-hole systems, compacting and imploding its false creations, reverting the energies back to the Source as 'space dust' fragmentation (back to pure unmodulated consciousness).

All life on this planet existing in a physical structure (body) is in a partially closed and finite system, which is considered normal; for example, man's unnatural carbon body (with impaired DNA—over 95% missing) of short finite life. In fact, all life on this planet survives by killing other life in order to survive. Science is partly looking at the unnatural creation aspects on this planet and galaxy, of which the latter has a dominant black-hole and is degenerating, reinforcing science's belief in finite closed systems and the resulting laws, such as the conservation of energy. In a sense, one might say science portrays a more degraded and detrimental creation than the

Metatronic; at least the latter recognises a perpetual source and that its survival (the Metatronic's) can be extended—beyond just the physical; by stealing energy.

Notes

1. Book: *Voyagers*, volume 2 by A. Deane.

2. Ibid.

APPENDIX C

SUBJECTIVE/OBJECTIVE VIEW OF CREATION

Without the subjective (the observer) there is nothing.

Figure 25 can be studied and used in relationship to the text in the book which deals with creation. The holograph and its fractal distribution is an interface energy configuration to express unity and separateness as it translates creation's unity and infinite nonlinearity into separateness and space-time linearity. As per the holographic structure there is one self manifesting in many parts and playing many roles, with those parts given restrictions on knowing, such as two parts not knowing one another but as we move up the scale, subjective/objective, the awareness of the other side increases.

In quantum reduction, the interaction of the observer, in effect, converts a degree of subjectivity into objectivity, the material environment. If the subjectivity has too high an order for the observing awareness or extent of 'information' it will quantum reduce into objective factors of the environment and form part of the overall unconsciousness (going up to the highest level). Appendix D presents this in a different form.

Science endeavours to remove all subjectivity. As we can see, at E, there can be no existence, only groupings of infinite infinitesimally small points (theoretical) or small particles (actual). At A the subjectivity has no separation, no parts, no unconsciousness. The more we increase the objectivity, going down the scale, the more we increase unconsciousness, or the degree of 'not-know'.

The Emerging New Science

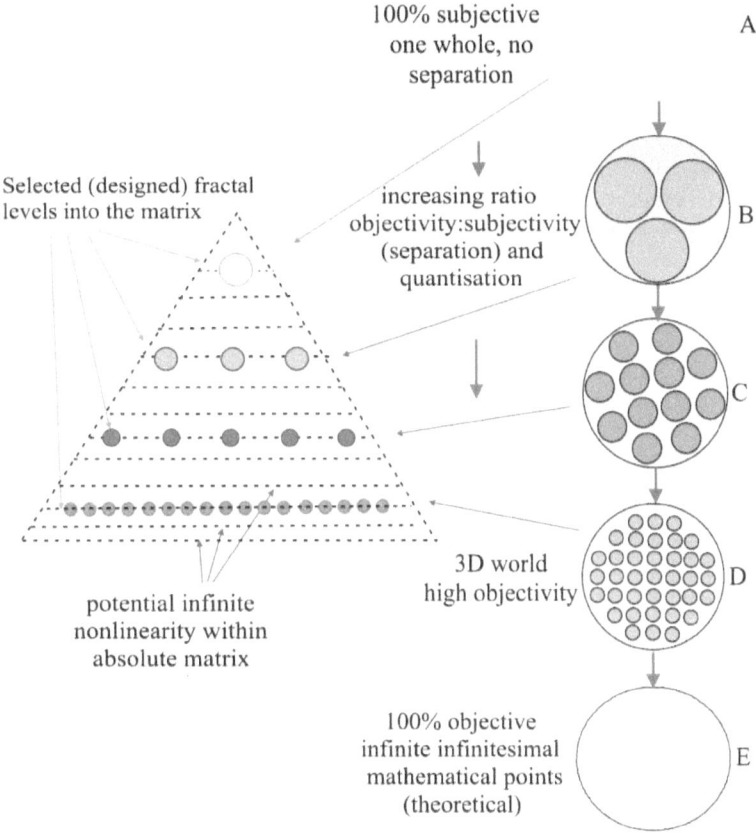

FIGURE 25: Creation concepts

Nevertheless these are fundamentally constructive features for the exploration of consciousness with different degrees of freedom.

We have argued that the objective universe is present but at the primary level of causation it is all mind—equally observers and observed with apparent separateness and objectivity. When the universe/environment is sufficiently unconscious (which also means objective, such as ours), it appears to actually function (relative to our limitation) more objectively. But we are influencing it more than we realise—the highest aspect of our consciousness is operating both sides (observer/observed). However, even at our level of high objectivity it is our interface with it that determines the view—such as a solid particle view.

If we could focus on the wave nature underlying our 3D view, indicated by quantum theory, hidden truths would be perceived much more easily. An advanced highly-computerised scalar-electromagnetic oscilloscope could reveal each person's identity spectrum, as briefly mentioned earlier.

APPENDIX D

EXISTENCE AND EVOLUTION, QUANTUM REDUCTION AND REGENERATION

The processes of ascension

We might conclude that the New Science could successfully be based within quantum theory, but the latter needs extending greatly—Einstein was correct when he stated that quantum mechanics was not complete.

We have seen that the nature of evolution is more quantum physics than any other branch of science. We have a fractal holographic dimensional structure of orders, within which the processes of quantum reduction and quantum regeneration take place—see Figure 26. The diagram is, of course, ludicrously simplified, showing only four levels or orders. All the levels (wave patterns) below the top, A, are infolded into A—we have separated them out for clarity. Nevertheless these fractal levels of wave-pattern universes exist separately as their design context dictates. It simply means that the selection arrows could be directed at A; it would amount to the same thing.

We see that the fractal levels provide the major fixed divisions in this hierarchy of orders. The divisions, however, have large gaps in them—jumps in the gradient of orders. Within these levels we have randomness and many possibilities, and the background fractal-matrix gradient.[1] Note that the fixed fractal levels have contextual zeros; meaning relative.

The evolving individual, however, provides the gradient —fills in the gap, so to speak. Consider an analogy: the ground-floor worker gradually acquires the qualification bit by bit to achieve the jump from the ground floor to, say, manager level. In life, this is the accretion of frequencies from the background, unified quantum field, plus the ordering and integration of these frequencies and frequency patterns into the individual's personality, leading to the required qualification to achieve the quantum leap to the next level.

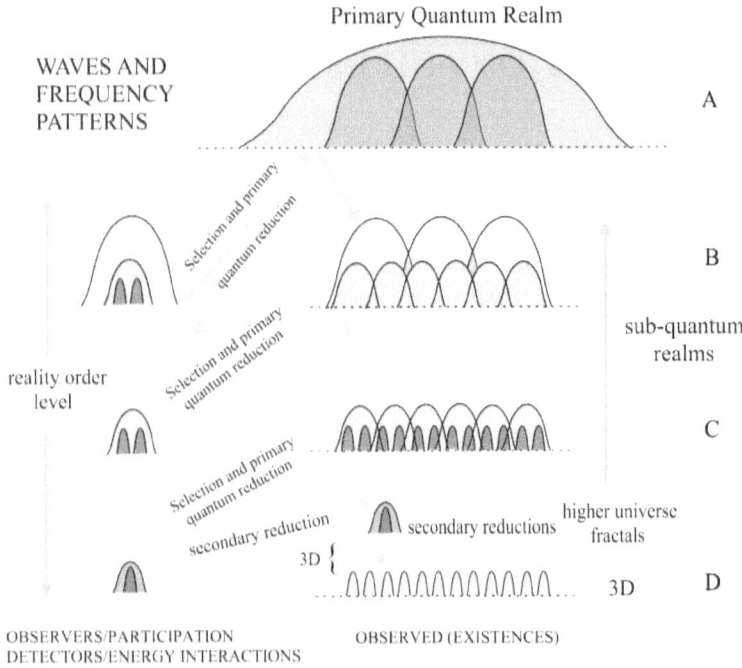

FIGURE 26

Thus the personality has his or her own frequency-spectrum identity and by means of the physics of wave patterns this energy structure, consisting of mind and consciousness, is drawn up or down the dimensional structure. Upwards requires greater coherence, downwards fragmentation and decoherence —the latter automatically drops in frequencies.[2]

The Emerging New Science

Amongst the randomness and disorder between fractal divisions, but more so at the lower levels, there are thus fractal orders within orders (recall that the basic hardware is always spheres within spheres—higher-dimensional vortices). As we note from Figure 26 the entity or device interacting with the universe receives back from the universe its own reality level. It is all about orders interacting with orders and the subsequent quantum reduction of higher orders. If this was all, then one would ask, How can one evolve (to a higher order)? If the observer is an artificial machine, the answer is, it can't. A human, however, can quantum regenerate elements from the environment. Note that relative to consciousness, the mind structure itself is, in effect, a mentally-empirical environment. By putting order into parts (for example, resonance) a greater (carrier) wave function is quantum regenerated, corresponding to a higher position in the scale. But where does this higher order come from if the observer resonates with and draws in its own reality level? Notice that it is not 'generate', but 're-generate'.

No matter how obstinate science is in its determination to establish a random beginning to creation, or achieving a higher order from a lower order, it simply can't be done; there must be initial, existing guide lines. In this case, the human has higher aspects of consciousness, higher orders, all the way to the top (which is 100% subjective). The observer's basic consciousness, without mind aspects, that is, particles, waves, is a nonquantifiable portion of the Absolute beyond space-time, particle and waves. Thus the human's extended mind, of holographic and fractally-organised particles and waves can always go one higher than its immediately-observed universe environment. This means people can put higher orders into lower orders of their immediate environment as they quantum regenerate to the higher order, drawing their mind/consciousness up to the next higher level (with the main effect usually after death of the physical body and more completely in entering the next life).

The human is a very complex entity; what about the basic physics of evolution at, say, the particle level? We have briefly explained this previously. Does the particle evolve? If so, how? When it is functioning naturally without mutations (DNA), or artificial and laboratory contrivances, the particle merges with its antiparticle (and not suffer the death of annihilation) to create a single higher-quantum state particle that instantly appears in the next higher-order fractal level, dividing again into particle and antiparticle in a new frequency-pattern relationship (in a different universe level; where the relationship between the particles is more integrated and of higher frequency). This continues up the levels to total coherence/integration in the unified field of the Absolute.

Thus the particle/antiparticle polarity in 3D is within, or carried by, the polarity of the next level, and so on, that is, polarities within polarities up the hierarchy, until 100% subjectivity is reached of no separation and no polarity.

Notes

1. www.nhbeyondduality.org.uk. Section: *The Fractal Matrix*.

2. Why does the frequency decrease in fragmentation of energies? The greater the wholeness of a coherent energy, the higher the frequency—which is rate of information and a feature of intelligence. Vice versa, fragmentation results in smaller and incomplete parts (even if small coherent regions), which are generally carried (controlled) by the higher-fractal coherent levels. The greater whole is the master control; thus the higher rate must be present (in the on/off process) whenever the lower rate is present. Furthermore, the cross-section, or interface, between the poles of the oscillation for the greater whole will have greater pressure (greater interface 'area') and thus will oscillate more rapidly.

APPENDIX E

OVERCOMING THE VELOCITY-OF-LIGHT LIMITATION

Relativity is a good representation of the illusions of the third dimension.

Relativity measurements have shown that the velocity of light is always a constant and that it is the upper limit to the velocity of a body—with the actual limit (of the velocity of light) not unattainable. We are particularly interested in what really happens to the body as its speed approaches that of light—how can the spacecraft travel (even) faster than the speed of light?

Light was first postulated to be travelling through a medium, named the aether. If the Earth moved through the aether, and light, say, from a distant star transmitted through the aether, then a difference in velocity of light would be expected between the measurement made when the Earth was moving towards the star and when it was moving away (opposite side of Earth's orbit round the Sun). Even with very high accuracy no difference was found. Science made the assumption that the aether would behave as a static medium, but further it then eliminated the aether altogether.

If we imagine the Earth as the centre of the dual-vortex system, that is, a 4D version of that in Figure 4, giving a spiralling-in effect in a spherical manner and affecting space (resulting in a similar effect to Einstein's space-curvature notion of general relativity theory), we might realise that the Earth would take the vortex wave pressure with it as it moved, and no aether drift would be measured—it would make no difference if there was a changing relative velocity between Earth and the incoming light, it wouldn't be detected because the light had to go through the vortex that was moving with Earth. Thus no matter in which direction Earth was moving, at whatever relative velocity, the incoming light would be constant relative to the vortex. In fact the light would accelerate into the vortex, giving rise to the red-shift mentioned in the Big Bang section. As light travels throughout the galaxy its velocity would be expected to vary considerably as it accelerated and decelerated through the vortex of the galaxy and within this, the vortices of stars and planets.

If one further attempted to measure velocity-of-light variations within the vortex, say, on Earth, that is, locally, owing to the experimental limitations and the relative velocities tested between the observer and observed, different frequencies (rates of information of space-time), we again may find the appearance of the constancy.

It was concluded that the velocity of light was always constant no matter what the speed of the observer was with the measuring device. But with the vortex explanation it doesn't have to be constant; if this was the case it would have a drastic effect on a large section of physics. The measured value of the light velocity will always appear to be constant if the aether vortex is creating the matter of the body with the body at the centre of the vortex. Out in space the velocity would vary. Once science establishes the (illusions of) velocity of light as universally fixed then appropriate distortions of space, time and mass have to be contrived to preserve the constancy of the light velocity. Then the third-dimensional limitations of the scientific instruments will confirm this.

By fixating the light velocity as a fundamental constant, it acts like the fulcrum of, say, a see saw—a zero point from which measurements are referenced. This zero point is relative to the third dimension and the latter's frequency spectrum or rate of information, or rate of creation of space. But our 3D is 'carried' by the next fractal level of space-time (which will have hyperspace properties) and a new 'zero'.

Here we are specifically interested in the mass-distortion illusion of matter when approaching the speed of light, which we shall see is clarified by the rate of creation of space. Relativity theory indicated that mass distortion effects of a body will occur at high speed, approaching light velocity, and that at the theoretical speed of light the body's mass will become infinite. Let's analyse this.

Special relativity says that as we push a body close to the speed of light its mass increases rapidly in the higher range up to infinite mass at the light velocity. Now we must admit that particles have since been accelerated by magnetic fields up to close to the light velocity and the force required to do this increases exponentially, which projects to infinity at the light velocity. But let's look at the assumptions here. Yes, the force applied is increasing as per relativity, so therefore the resistance is also increasing proportionately. Everything is fine at this point. But we might ask, is this resistance the normal one, which has increased exponentially, or is it an extra resistance factor which has emerged? Science assumes it is the normal resistance due to inertia; a big, unjustified assumption. Thus it is decided that the inertia also increases exponentially, and if it does we know that this means the mass has increased likewise. This is the thinking.

Science recognises that our material existence is in 3D, and that these three dimensions are orthogonal (at right angles to each other). However, there is no detection by science of the spectrum of frequencies that these dimensions encompass. Scientific instruments can't detect them for the simple reason that the instruments themselves and the brains and physical senses of the observers are manifestations from the same frequency spectrum (a machine can't evaluate a machine of the same order). The vortices of all bodies: planets, stars, galaxies, are creating space-time as well as mass, and their waves interfere, causing criss-crossing of energy lines and a myriad of mini-black and mini-white holes, appearing as micro-particles and virtual particles; nodes, in fact. There are thus countless vortices, large and small, forming space; it is a mass of oscillations, in effect, primary and secondary nodes. An oscillation has a frequency 'swinging' from positive through neutral (zero value) to negative. Thus a basic oscillation is like a pulse switching on and off.

Keeping the above in mind, we can now see that an object being accelerated artificially within this field of activity and is out of phase with it, that is, its own oscillations (of atoms) are not coherent with space-time, when travelling at high velocity, if its linear speed becomes of the order of magnitude of the rate of motion of the space oscillations, it is held back during the 'off' state of the oscillations. That is, it has to 'wait' for the space to be created in front of it. There is thus a dragging effect on its motion, characteristic of resistance. Since these first-order space oscillations determine the velocity of light, the body will meet with impossible 'resistance' at the speed of light.

This would be the additional 'resistance' (to the inertia) that impedes the body through space as it nears the velocity of light. It is the relatively low rate of information of 3D space that prevents the body going faster. But this will be a 3D fractal level caused by the mini-oscillations (vortices). There will also be larger oscillations from larger entities, such as planets, stars, galaxies, etc. These are greater coherent states/nodes with higher frequency, giving rise to inner layers of hyperspace. If the body's complete structure itself is made coherent (entrainment of atomic oscillations that raises its frequency) it could theoretically match and resonate with this next hyperspace fractal level of higher frequency and then be subjected to a new and higher upper limit to the light velocity—meaning a greater speed of light in higher (hyper) space. And so on to the next level. However, this body is still obeying Newton's laws but is only experiencing normal resistance now due to genuine inertia alone.

The above is a relatively crude system compared with the one explained previously (Figure 5), in which the spacecraft is coherent with space-time but is not being pushed through space. This craft's coherent oscillation, which could be controlled in frequency value, will be put into phase with the oscillations of space-time at the different fractal levels, described above. A less advanced version of this craft might only be able to match the lower range (our first-order light velocity) but nevertheless attain light speed effortlessly with no inertia problems. If it matched the galaxy frequency (enormously high) it would probably travel thousands of times faster than the light velocity; with no inertia.

There is an interesting extension here, which will never be believed by relativists, biologists and physiologists. The above relativistic data relates closely to the learning-pattern properties described earlier. At any level of development of the learning pattern (for physical movements) there is a particular value of the rate of information. Thus just as the craft pushes its way through space at relativistic speeds, experiences a dragging effect, so the learning pattern of the mind-computer system causes an apparent resistance in muscular movement, which science thinks is physiological.[1]

A further interesting point to be made here is regarding the learning-pattern hardware within the quantum-field systems around the muscles and joints of the physiological body, during making skilful physical movements, particularly at high speed and in complex coordination. The field system, in addition to giving a boost to the power (which is dynamic strength), also acts to some degree in the manner of the coherent body, bypassing inertia. For example, the arm of, say, a concert pianist, when playing quickly to invoke considerable inertia, is not experiencing the full value of the expected Newtonian forces and inertia. [The author, having applied this knowledge, has acquired extremely low resistance in joints/muscles (age 80). Scales on a keyboard can be played faster than a concert pianist, and a further example was a sprint test about a year ago (across a 50-metre lawn), not having run for about 60 years, and finding it, at the first attempt, as easy as when a teenager, and probably as fast.]

Moreover, if we extend this non-Newtonian information to the subject of insect flight we will see that accompanying the physical mechanics of the insects wings (muscles and skeletal frame) there is a coherent oscillating quantum field system in which the atomic oscillations of the wings are entrained to quantum regenerate a single oscillation encompassing the wings, and there will be cancellation of inertia of the wings. Thus similar to the spacecraft's one-bodied coherence with space-time the insect's oscillations and wing mass form a single coherent system. Otherwise some insects would never be able to beat their wings hundreds of times per second.[2]

Notes

1. www.nhbeyondduality.org.uk. Articles on skills, and book, *The Attainment of Superior Physical Abilities* by N. Huntley.

2. Ibid. Article: *New Look at the Flight Mechanics of Insects*.

BIBLIOGRAPHY

Bearden, T. E. Solutions to Tesla's Secrets and the Soviet Tesla Weapons. Milibrae, California: Tesla Book Company, 1981.

Bohm, D. Wholeness and the Implicate Order. London: Rootlege & Kegan Paul, 1980.

Capra, F. The Tao of Physics. New York: Bantam Books, 1977.

Chown M. Quantum Theory Cannot Hurt You. London: Faber & Faber Ltd, 2007.

Cooper, A. P. Our Ultimate Reality. Ultimate Reality Publishing, 2007.

Davies, P. Other Worlds. New York: Simon & Schuster, 1980.

Davies, P. The Cosmic Blueprint. New York: Simon & Schuster, 1980.

Deane, A. Voyagers: Secrets of Amenti, volume 2. Columbus, NC: Granite Publishing, 2002.

Fara, P. Science: A Four Thousand Year History. London: Oxford University Press, 2009.

Feynman, P. R. The Character of Physical Law. London: Penguin Books, 1965.

Goswami, A. The Visionary Window. Illinois: Theosophical Publishing House, 2000.

Gribbens J. Science: A History. London: Penguin Books, 2002.

Hawking, S. A Brief History of Time. New York: Bantam Books, 1988.

Herbert, N. Quantum Reality: Beyond the New Physics. New York: Anchor Press/Doubleday, 1985.

Huntley, N. The Attainment of Superior Physical Abilities and the New Science of Body Motion. USA: Xlibris Corporation, 2005.

Huntley, N. The Original Great Pyramid and Future Science. London: Author House, 2011.

Huntley, N. A Paradigm for Consciousness (parapsychology dissertation). St. John's University, USA, 1987.

Kumar, M. Quantum: Einstein, Bohr, The Great Debate about the Nature of Reality. London: Icon Books Ltd, 2009.

Norman, E. L. The Infinite Concept of Cosmic Creation. El Cajon, California: Unarius, Science of Life, 1970.

Ouspensky, P. D. The New Model of the Universe. New York: Random House.

Russell, W. The Secret of Light. Virginia: University of Science and Philosophy, 1974.

Rucker, R B. Speculations on the Fourth Dimension. New York: Dover Publications.

Talbot, M. Beyond the Quantum. London: Bantam Books, 1987.

Toben, R. Space, Time and Beyond. Toronto: Irwin Company Ltd., 1975.

Vedral, V. Decoding Reality. Oxford: University Press, 2010.

Wolf, F.A. Parallel Universes. New York: Simon & Schuster, 1990.

Wolf, M. The Catches of Heaven. Pittsburgh, PA: Dorrance Publishing, 1996.

INDEX

A

action concepts 234, 247
aether 40, 51, 89, 153, 295
aether vortex 296
antigravity 32, 35, 48, 161
antimatter 152
antiparticles 87, 96, 152, 157, 195
Anunnaki 27
Armageddon 282
art and music 78, 143, 239, 265
ascension 63, 67
Aspect 50
astral 86, 89, 158, 194, 207, 282
asymmetric 72, 74
atheist 66, 102, 255

B

Bible 26, 27, 125
binary system 181, 186, 278, 282
black-hole 24, 87, 96, 108, 286
Bohm, D. 18, 20, 51, 59
Bohr, Neils 138, 153, 198
broken symmetry 74
Brown, William 220
bury 28

C

carbon body 286

cells 27, 105, 106, 110, 116, 273
Charon 282
Christianity 66, 75
Clark, Terry 113, 128
coherent quantum state 188, 225, 244, 254
Collective 47, 76, 143, 208, 275
combustion engine 14
contextual 11, 16, 30, 33, 38, 46, 152, 168, 198, 221, 243
Copenhagen Interpretation 50, 53, 185, 203, 207, 219, 238
cosmic microwave 94
creation of space 296
cremate 28, 29
Crookes, William 86
crop circles 273
curved space 41

D

Darwin 13, 83, 90, 102, 168
Deane, Ashayana 282, 285, 301
Denton, M. 106
dictyostelium 156
DNA 16, 26, 29 56, 79, 86, 1020, 273, 285, 293
junk 16, 26, 31, 102

dualism 98, 133, 169, 173
duality 145, 146, 169

E
Edison 47
ego 12, 31, 47, 53, 68, 75, 76, 104, 122
ego-self 68
Einstein 41, 107, 138, 153, 171, 179, 236, 239
Einstein-Bohr debate 204
empiricism 11
   mental 32
   physical 32
ESP 77
Everett, Hugh 208
expansion 14, 32
experimental psychologists 148
extraterrestrials 282

F
Fall of Man 164
fibonacci series 286
flatland 19
flatness problem 94
force field 245
   electric field 46, 71, 197
Fourier analysis 217, 223
fractal matrix 220, 246
free fall 160, 162, 197, 238, 245
Freedom Teachings 282, 285
fusion 152

G
galaxy 24, 42
   Andromeda 25
genes 106
geometry 19, 27, 31, 45, 155, 167, 240, 252
   absolute system 155, 241, 273, 279
God 65, 66, 274
   external 66
God Source 66, 88
golden mean 286
gravitational force 160, 195, 235
gravity 32, 40, 46, 161, 235
Great Pyramid 27, 285, 302
Guardians 282-286

H
harmonic science 44, 48, 54, 74, 235
harmonic systems 70
Hawking, Stephen 98
Heaven 89, 303
heretical 35, 48
holistic 12, 174, 185, 194, 229, 235
holographic 20, 33, 36, 34, 44, 51, 128, 183, 222, 247, 275
holographic amplification 231
holographic fractal system 17, 32, 39, 96, 145, 181
holy 108, 264
horizon problem 94
hyperspace fractal level 298

I
illusion 45, 52, 85, 98
inertia 297, 299
infinitely nonlinear 121, 131, 152, 173, 177, 214
Inflationary Model 96
intuition 32, 45, 93, 144, 229, 243

J
Jesus Christ 19, 63, 80, 102, 284

K
karma 42, 102, 123, 145, 146, 164
Keylontic Science 286
Kingdom 19
    Father 19
    within 19
Krist 285
Krystal spiral 287

L
learning pattern 36, 46, 56, 115, 224, 229, 232
    aging 231
    habit pattern 231
    kinaesthetic sense 232
Lee, Chang and Wu 74
Leibnitz 20
Lodge, Oliver 86
Luciferian 26

M
Maldek 26, 282
many worlds interpretation 208
Maxwell 51, 74
Metatronic 284, 286
Milky Way 24, 25, 42, 108
mini-black holes 87, 95, 195, 297
multiverse 21, 32, 33, 37, 39, 121, 128-130, 141, 148

N
Newton 15, 41, 46, 51, 70, 161
Newtonian 136, 160, 162, 163,
Newton's laws 162, 197, 235, 245
    bypassed 235, 161, 162, 163
Nibiru 26, 282
    diodic crystal 26
    Nibiruian Battlestar 282, 283
Nobel Prize 74
nonlocal 50, 108, 135, 173, 187, 194, 200
nothingness 85-87, 96, 125, 167, 175, 196

O
observer/observed 12, 15
    relationship 12, 207, 210
    oscilloscope 290
        scalar electromagnetic 290

Ouspensky 20, 302
override systems 42

P
parasitic evolution 47, 58, 108, 284, 286
particle 96, 152, 195
    evolution 97
Pavlov 108
Penrose, Roger 210
Philadelphia experiment 92, 119
planet X 282
progress 13
    combustion engine 14
propulsion 54

Q
quanta 181, 193, 196, 208
quantum action 46, 160-162, 208
quantum evolution 110
quantum foam 20
quantum randomness 182, 201
quantum reduction 38, 48, 58, 69, 125, 141, 170, 176, 288, 291
    secondary 197, 210
    objective reduction 198, 210
    subjective reduction 198, 210
quantum regeneration 32, 45, 54, 122, 141, 198, 201
quantum states 33, 46, 52, 112, 113, 117, 127

quantum uncertainty 178, 207, 257
quaternion 51

R
red shift 94
Relativity 41, 49, 107, 295-297
religion 12, 19, 32, 54, 65, 89, 102, 284
re-Legions 136
    Twelve Legions of the Christ 136
rocket 54, 161, 163
Roll, Michael 86, 117

S
sacred science 54
Sarfatti, J. 20
Satan 284
scalar 26, 46, 77, 144
scalar field 161, 162, 196, 197, 233, 246
scalar generator 253
Schrodinger 34, 53, 187, 196
scientific methodology 21, 24, 69, 81, 91, 183
sentience 34, 53, 88, 112, 125, 126, 133, 269
singularity 95, 96, 98
    portal 95
Sirian Council 27
soul 103, 130, 158, 231, 266
    level 143
Source 19, 51, 65, 66, 88

spacecraft 163, 188, 235
    one-bodied 298, 299
    vortex 55
spiritually 85
stacking order 224, 225
stealth 92
Stonehenge 26, 27
string theories 49, 98
subjective/objective levels
    16, 134
subjectivity/objectivity
    142, 207
superhologram 20

T
technologies 14, 28, 44, 54,
    55, 68-70, 162, 163,
    235, 246
    one-bodied system
    55
teleportation 185, 251
    object 'remoulded' 255
television 11, 12, 21, 52,
    62, 83
Tesla, Nikola 47, 301
turbine 47, 72

U
unmanifest potential 112,
    121, 139, 144

V
vacuum 40, 71, 86, 96, 195
virtual particles 71-74, 87,
    96, 111, 195, 196
    boson 180, 195
    gluon 195
    graviton 161, 195
Von Neumann 100

vortex 33, 36, 40, 41, 55,
    95-96, 150, 240

W
waterfall 71, 73
    analogy 71
Wheeler, J. 92
white-hole 24, 40
Wormwood 26-28, 282

Z
zero 16, 45, 69, 70, 109,
    196
zero-point energy 195, 196

www.ingramcontent.com/pod-product-compliance
Lightning Source LLC
Chambersburg PA
CBHW021351210526
45463CB00001B/59